Hidden In Plain Sight

Andrew Thomas studied physics in the James Clerk Maxwell Building in Edinburgh University, and received his doctorate from Swansea University in 1992.

He is the author of the *What Is Reality?* website (www.whatisreality.co.uk), one of the most popular websites dealing with questions of the fundamentals of physics. It has been called "The best online introduction to quantum theory".

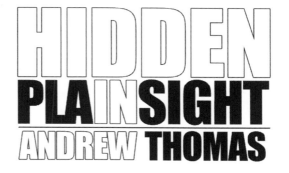

HIDDEN PLAIN IN SIGHT

ANDREW THOMAS

DEDICATION

To Mum and Dad

CONTENTS

PREFACE

For the last seven years I have been running a website on fundamental physics (www.whatisreality.co.uk). I am pleased to say that over that period the website has grown to become one of the most popular physics sites on the web.

The seven years working on the site have provided me with the opportunity to consider some of the most challenging questions about fundamental reality. Some avenues which initially appeared promising turned out to be blind alleys, while other seemingly innocuous principles turned out to be surprisingly fruitful. This book presents some of the most important insights I gained over those seven years.

I am pleased to say the website has gained high praise from some of the most famous names in physics, but I have been more heartened by some of the comments left by casual visitors to my site, especially those who have appreciated the site and felt I managed to make physics exciting, entertaining, and easy to understand. Hopefully this book can continue that trend.

Andrew Thomas (hiddeninplainsightbook@gmail.com)
Swansea, UK,
2012

u·ni·verse (yo͞o′nə-vûrs′)
n.
   1. The universe is one thing,
   the only thing that exists.

1

UNIFICATION

In our modern world, we have grown used to ever-increasing complexity. We perpetually have to upgrade our mobile phones to the latest version 5.0 with new features we never knew we wanted. We have to buy the latest television with hard-disk recording built-in, and a baffling remote control to match. It feels like it is just a matter of time before the internet becomes so complicated that it develops self-awareness, calls itself "SkyNet", and launches a thermonuclear war against humanity.

Complexity is seen as a virtue, a selling-point. Extra features are seen as a good thing. Are you not confused enough yet?

I want to strip my life back to simplicity. I have cancelled my Facebook account. I have organised my apartment, stripped it down to the basics and discarded all unwanted junk. By removing distractions, I hope it is going to allow me to concentrate. When you feel the complexities of life are getting you down, you have to strip your life back to simplicity.

We tend to think of simple things as more beautiful. They are more economical, with less waste, less superfluous clutter. If we clear our heads, we can see things more clearly. If we clear our heads, we can see the truth.

Nature also likes simplicity. Nature abhors waste, both in materials and in energy. Everything in Nature is done for a good reason. We see beauty in simplicity because we can see its underlying efficiency. We might recoil from seeing an ostentatious sofa because so much of its covering serves no purpose, the inefficiency of its design being antithetical to the beauty of Nature. However, when we see a sleek, lean, minimally designed chair we are instantly attracted to its spartan beauty (perhaps until we have to sit on it for a couple of hours).

Due to Nature's close affinity with simplicity, the advance of fundamental physics over the centuries could be seen very much as the pursuit of simplicity. As physics has progressed it has been found that more elegant, simpler theories have been found to be the best match for reality. It is very important to remember this when we do physics: ask yourself, are you making things simpler? If not, then think again. As Albert Einstein said: "Any intelligent fool can make things bigger and more complex. It takes a touch of genius – and a lot of courage – to move in the opposite direction."

In physics, one of the most successful ways of simplifying our explanations has been by reducing redundancy in those explanations. If we find we have two theories which explain two apparently different behaviours of Nature, we might find it possible to replace both those theories with a single theory which still manages to explain both of those behaviours. When this happens, the process is called *unification*. The resultant unified theory will be simpler than either of the two previous theories, as it reveals a deeper, underlying truth.

This process of unification always represents a landmark in physics. It is like drawing a line in the sand. It shows we have reached a new level of understanding – we will never have to go back to our previous level of ignorance. Unification is, in many ways, the ultimate goal of physics. If physics is ever to provide a complete understanding of the universe, then the last act which will be performed by the very final physicist will be an act of unification.

This book has a single goal, though it is highly ambitious. The aim of this book is to reveal a link – possibly a unification – between relativity and quantum mechanics at the deepest possible level: the fundamental level. This link is based on an extremely simple principle, a simple idea.

Simplicity is good. I like simplicity.

What unifies unifications?

As we start out on our quest, a good starting point might be to examine some of the major unifications in the history of physics. Maybe we can detect trends or similarities in their approach. Maybe we could adopt aspects of their approaches to aid our quest. With this in mind, what follows is an historical review of some of the major unifications in physics. Don't worry too much about following the details – any principles which are important for our discussion will be described in detail later in the book. For the time being, just get a feel for the history of unification.

We start in 1638 when Galileo, the man described as the father of modern science, published his *Discourse on Two New Sciences*. Galileo realised that if you are travelling at a constant speed in a constant direction (for example, being a passenger on a steady train), you feel as if you are stationary. Galileo had the insight that it was fundamentally impossible to distinguish the two situations of moving and being stationary. To be precise, Galileo stated that there was no experiment you could possibly perform which could detect if you were stationary or moving at a constant velocity. This unification of "being stationary" and "moving" stands as the first of the great unifications in physics.

In 1665, a 22-year-old Isaac Newton wondered if the force which pulled apples from a tree was the same force which held planets in orbit. Again, we see a simple, brilliant insight, an imaginative leap. From Newton's clear vision, a description of the force of gravity emerged as the unifying theory for two behaviours which were previously considered to be unrelated.

Another unification occurred in the nineteenth century by the man widely acknowledged to have been the greatest experimental physicist of all time. In 1831, at London's Royal Institution, Michael Faraday demonstrated that if you push a magnet through a coil of wire, an electric current flows. Conversely, if you pass an electric current through a wire it can deflect a nearby magnetic compass. From this, Faraday deduced that electric currents create magnetic fields, and moving magnetic fields create electric currents. Thus was *electromagnetism* discovered, unifying electricity and magnetism. However, Faraday was no mathematician, and he lacked the mathematical language to describe his discovery.

The next unification was revealed by the Scottish physicist James Clerk Maxwell who – unlike Faraday – **was** a mathematician. Maxwell took Faraday's results and constructed four equations which described the connection between electric and magnetic fields. Maxwell showed that an electric field would generate a magnetic field, and, conversely, a magnetic field would generate an electric field in an oscillatory fashion. The result was a self-sustaining electromagnetic wave which travelled through space. After calculating the speed of this wave, Maxwell found it to be the same as the speed of light, and thus had a brilliant insight: light is a form of electromagnetic wave! Hence, here was another unification, this time linking the field of optics with electromagnetism.

The next series of unifications happened in the twentieth century and was supplied by Albert Einstein, whose work we will be considering in detail later in this book. From a young age, Einstein had wondered what would happen if you moved so fast that you caught up with a light beam. It would seem that you would view a stationary wave of light. However, Maxwell's equations seemed to indicate a constant speed of light – regardless of the motion of the observer. Hence, Maxwell's result for the motion of light did not agree with Newton's laws of motion. Einstein managed to combine Maxwell's result with Newton's laws – but only by radically altering our concept of space and time. His resultant special theory of relativity unified not only electromagnetism with mechanics, but also space with time. Also, through his famous formula which arose from special relativity, $E = mc^2$, Einstein launched us into the atomic age by unifying mass with energy.

Einstein was on a roll at that point, and, ten years later, after a great deal of extremely hard work, Einstein struck again. Einstein had the brilliant insight – which he called the happiest thought of his life – that the force experienced by an observer undergoing constant acceleration was indistinguishable from the force of gravity. This represented another unification, this time between gravity and accelerated motion. Conversely, if an object was in free-fall it would not feel any force of gravity – it would be as if it was floating freely in space. This meant that gravity did not exist as an entity in its own right: gravity was the curvature of space. Objects travelling in a straight line in curved space appear to be drawn towards the centre of curvature, and this is interpreted as the force of gravity. The resultant theory – the general theory of relativity – unified gravity with the geometry of space.

The next unification occurred shortly after the discovery of relativity. The newly-developed theory of quantum mechanics was not consistent with special relativity as it only considered particles which moved very slowly compared with the speed of light. In 1927, Paul Dirac combined special relativity and quantum mechanics to create the first quantum field theory. This theory was further refined throughout the twentieth century to become the greatest triumph of physics in recent years as the foundation of the standard model of particle physics. The shy Dirac was awarded the Nobel Prize in 1933, and his first instinct was to turn it down as it would generate unwanted publicity for himself — until he was persuaded that turning down the prize would generate a lot more publicity.

The last unification to date occurred in the 1960s when physicists started to consider the symmetry between the particles associated with electromagnetism (the photon) and the weak nuclear force (called the W^+, W^-, and Z particles). Unification was achieved by importing an idea which was first used in studying the physics of metals and other solids: *symmetry breaking*. Physicists realised that all four particles would be massless (i.e., they were all symmetrical) at high energies (for example, the unimaginably high temperatures reached in the first moments after the big bang). However, as the energy decreased and the universe cooled, this symmetry would be broken and the particles would be free to take different values. The photon was left with no mass, but the W^+, W^-, and Z particles all collected mass. For this reason, the weak nuclear force can only operate at short ranges and is weak, whereas massless photons can bring us light from the stars. Hence, the two forces of electromagnetism and the weak nuclear force were, in fact, the same force behaving in two different ways at our everyday low energies. If it is sufficiently hot enough then the electromagnetic force and the weak force would merge into a single force. This was confirmed in the 1980s in CERN, the European laboratory for particle physics, when they recreated the temperature of the early universe and demonstrated the existence of the so-called *electroweak* force.

Considering general themes in all these approaches, one thing our historical study has revealed is that, in general, unification in physics is achieved by a single, imaginative idea. This might come from Newton pondering on an apple, or Einstein deciding that you could never catch up with a light beam: unification always requires one single, brilliant insight. Even in the case of the electroweak unification, with many physicists involved, it still required the inspired introduction of the symmetry breaking concept before unification could be achieved. It does not matter how much money or how many researchers you throw at the problem, unification is only ever achieved by a moment of inspiration. Unification is definitely not achieved by design-by-committee.

Also, the ideas by which unification is achieved are generally extremely simple ideas – anyone could understand these ideas. In fact, we could say that any unifying idea **has** to be simple. The idea of experiencing no gravity in a lift in free-fall, or catching up with a light beam, are simple ideas which anyone could understand. Similarly, Newton's idea connecting a falling apple with a planet held in orbit is a simple concept. It was as if these great unifiers picked-up on something which was under our noses all the time, something which was missed perhaps because it was too simple. Something hidden in plain sight.

The process of unification could be equated to a tree, with each theory being a leaf on the end of a twig. As we move back down the twig we eventually get to a junction point between two twigs: a point of unification. As we move further down, twigs merge into branches, and the branches merge into the trunk. As we continue to move further and further down the tree we find the number of branches gradually reduces as the process of unification reduces the number or redundant branches (theories). With each step, with each unification, our model of Nature becomes simpler.

The following diagram shows this tree of unification. The junction points of the branches represent points of unification:

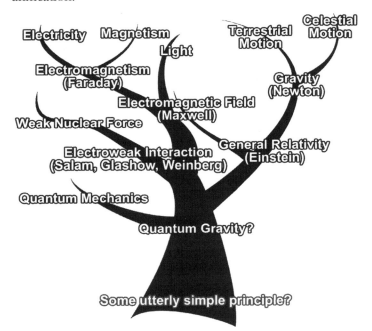

But what will we find at the base of the tree? What is holding up the entire structure? As our progress down the tree has seen our model and our theories getting simpler, you would expect that at the ultimate base of Nature, we will surely find a principle which is simple, beautiful, and elegant. As the great physicist John Wheeler said: "To my mind there must be, at the bottom of it all, not an equation, but an utterly simple idea. And to me that idea, when we finally discover it, will be so compelling, so inevitable, that we will say to one another: 'Oh, how beautiful. How could it have been otherwise?'"

Relativity and quantum mechanics

The quest for one particular unification is dominating current research. As we enter the twenty-first century, we find the same two theories have dominated physics for a hundred years: the theories of general relativity and quantum mechanics. Both theories have been astonishingly successful and have been tested to remarkable accuracy. General relativity is our best theory of gravity, accurately describing the motions of planets, stars, and galaxies, and has correctly predicted the existence of black holes. Quantum mechanics is the force behind the laser and the transistor, and is therefore the force behind the computer revolution. Quantum mechanics is generally regarded as the most successful theory in the history of science. Nobody argues with either theory – they are both undoubtedly correct

The only problem is, they have absolutely nothing in common.

General relativity is the theory of "big stuff", describing the motion of planets and galaxies. Quantum mechanics is the theory of "little stuff", describing the behaviour of sub-atomic particles. For the most part, physicists only have to use one or other of these theories, depending on the scale of their subject. For example, in order to study the motion of galaxies, they would use general relativity, in order to study the behaviour of particles, they would use quantum mechanics. So, because of the differences in the domain of application, this method of switching between the two theories – though not ideal – works perfectly well for most purposes.

However, we are interested in the quest to uncover the fundamental truth of Nature, and Nature surely does not use one theory for small objects and another theory for big objects. As far as Nature is concerned, objects are objects – the laws of physics apply equally to all objects. So at a deeper level, there must be a more elegant theory which applies to all objects, regardless of their size. It is the unearthing of this deeper theory – combining general relativity with quantum mechanics – which is the focus of much current research.

Einstein's theory of general relativity revealed that the very fabric of space itself can be curved in the presence of mass or energy. In the absence of mass or energy, space is "flat", it is uniform and without features. However, in the presence of mass or energy, the fabric of space is curved. This is a wonderful sweeping theory. One can imagine vast tracts of space being bent in a beautiful, smooth curve by a planet or star, like an interstellar ocean wave. And, just like an ocean wave, this glorious curve of space keeps moving and curving as the planets and stars are constantly in motion.

But what, exactly, is happening at the microscopic level during this curvature of space? What does space look like when we zoom in to examine this curvature in detail?

In the microscopic world, quantum mechanics takes over. Quantum mechanics predicts seething randomness at such small levels, with particles flashing into existence before their rapid annihilation. The resultant effervescent substance resembles foam rather than smooth space, and was given the name *spacetime foam* by John Wheeler. It will surely be a major challenge to analyze such a structure, if it even exists.

Theories which attempt to unify general relativity with quantum mechanics at this small scale are called theories of *quantum gravity*.

String theory and loop quantum gravity

The two main approaches to finding a model of quantum gravity are string theory and loop quantum gravity. These two approaches are opposites in many ways (loop quantum gravity emerges from general relativity, string theory has a closer connection to quantum mechanics), but very similar in one important way. Both string theory and loop quantum gravity are based on the principle that the best way to analyse Nature is to discover the *structure* of the smallest components of reality.

String theory was first proposed by Yoichiro Nambu in 1970 while at the University of Chicago. When first introduced, the theory had more modest goals than it has now. The theory has been considerably expanded over the years, but the fundamental principles have been retained.

String theory considers particles as being composed of microscopic strings:

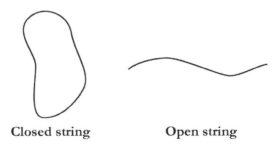

Closed string　　　　　**Open string**

As a closed string moves through space and time, it sweeps out the shape of a cylinder. These cylinder shapes can simplify our understanding of the interactions between particles. As an example, the following diagram shows the interaction between an electron and a photon:

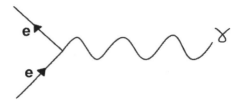

The shape of this interaction between elementary particles can then be viewed as a single string cylinder pinching itself and separating into two string cylinders:

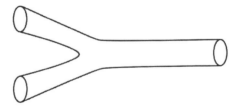

The great attraction for theorists is that this stringy cylinder eliminates the infinities associated in dealing with the interactions of point particles. As can be seen in the two diagrams above, the particle interaction can be replaced by the much smoother string cylinder representation.

As we take successive snapshots in time, cross-sections through the string cylinder appear as microscopic loops which join and split in a smooth manner:

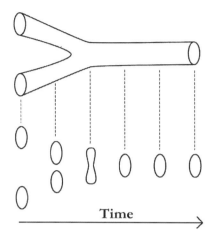

Time

If these microscopic loops are sufficiently small, they can appear just like point particles. So the apparent interaction of point particles could actually be the splitting and joining of microscopic string loops.

Just as a guitar string can vibrate in many different ways to produce different notes, so can a loop of string have different modes of vibration to produce different types of particles. For a loop, a whole number of vibrations must fit around the string. Higher notes require more energy, so the particles would have greater mass. In this way, different particles – such as electrons and quarks – are just different vibrations of one of these tiny loops of string.

Much is made of the fact that string theory predicts a particle which appears to have the characteristics of the particle which is expected to carry the gravitational field. Even though such a particle has never been detected experimentally, it has already been given a name: the *graviton*. Such a particle would have to treat all objects in the same way – just as gravity is an attractive force for all objects. The fact that such a particle emerges naturally from string theory is often taken to mean that in string theory "gravity comes free". String theory's prediction of gravitons is one of the main selling-points of the approach.

This all sounds well and good, until you realise that string theory makes another prediction which is not so welcome, the prediction that space should have 10 or 11 dimensions. This is a blow, as we only see three spatial dimensions and one time dimension. This has forced the ingenuity of string theorists to come to the fore, as ways of tucking-up the extra dimensions into exotic structures have been suggested. However, having to resort to these kind of techniques is not really what you want in your simple, fundamental theory.

String theory is almost the only game in town when it comes to current theories of quantum gravity. However, there is one more proposal being worked on by a smaller community of researchers. *Loop quantum gravity* attempts to unify quantum mechanics and general relativity by applying the standard technique of "quantization" to Einstein's equations for general relativity. The key figures in this field are Lee Smolin of the Perimeter Institute near Toronto, Carlo Rovelli of the University of the Mediterranean in Marseille, and Abhay Ashketar of Pennsylvania State University. We will be hearing some of these names throughout this book.

Loop quantum gravity's initial aims were less ambitious than those of string theory, being purely a theory of quantum gravity, with no attempt being made to predict the properties of elementary particles. However, this has changed in the latest version of the theory, which now does, indeed, predict the elementary particles. So the latest version of the theory is, like string theory, presented as a fully-fledged *theory of everything*. The main prediction of the theory is that space is not continuous but is instead quantized into discrete chunks: "atoms" of space. These chunks are the smallest possible units of volume, each approximately 10^{-99} cubic centimetres.

Neither of the approaches of string theory nor loop quantum gravity seems to be enjoying unqualified success at the moment, with progress being slow on all fronts. String theory has been getting a bad press as the theory which makes no testable predictions. Loop quantum gravity, however, boldly proposed a testable prediction that high-energy photons should travel slightly slower than low-energy photons, and this difference becomes measurable over long distances. However, in 2009, a gamma ray burst from deep space was analyzed and no speed difference could be detected.[1]

There is a certain competitiveness/animosity between researchers in these two distinct fields, both groups being convinced of the correctness of their approach and the failings of the other approach. Such competitiveness should be healthy in theory, but it has not appeared to work out that way. This is not a golden age for physics.

A personal criticism of mine − which applies to both string theory and loop quantum gravity − is that neither theory has anything to say about the foundational problems of quantum mechanics. The foundational problems raise questions about the nature of reality at a very low level − before it is even observed. It is hard to even imagine what form reality takes before it is observed, so this might be considered more of a philosophical question by most scientists, and is sadly ignored by researchers, but it is still a problem at the very heart of reality which urgently requires a solution. We will be considering the foundational problems in Chapter Six which is dedicated to the question of

---

[1] Although this prediction of loop quantum gravity was disproved by experiment, Lee Smolin was not disheartened. In fact, he has stated that he is satisfied that this approach represents "doing real science" (the implication being that string theory is not real science).

quantum reality. No matter how long they are ignored, the foundational problems will not go away and will have to be tackled at some point. One of the proposals of this book is that a true linkage of quantum mechanics and relativity can only be properly achieved by obtaining a greater understanding of the foundational level.

The problem with structure

It was the Greek philosopher Democritus who first suggested that matter might be composed of indivisible "atoms" (meaning "uncuttable" in Greek). Democritus believed the properties of the atoms were determined by their shape, so water atoms were smooth and slippery, whereas iron atoms had hooks to bond closely together. His views might sound misguided, but, on the basis of what we now know about chemical bonding, it is actually a remarkably good analogy to reality.

Throughout the nineteenth century, successive experiments appeared to confirm the atomic nature of matter, until it became an accepted fact. But were the particles truly indivisible? In 1897, the physicist J.J. Thomson discovered the electron, revealing that there were smaller particles within the atom. In 1909, Ernest Rutherford directed alpha radiation at some thin gold foil, and found it was deflected. Rutherford interpreted this result as showing that there was a solid, positively-charged nucleus at the centre of every atom, surrounded by negative-charged electrons.

The continuing subdivision of particles continued through the twentieth century with the discovery that protons and neutrons were each composed of three particles called *quarks*. Even though quarks are considered to be elementary particles, if history teaches us anything it is that particles which were once considered indivisible later reveal themselves to be composed of even more fundamental particles.

In answer to the question "What is a particle?", researchers in string theory would say elementary particles are actually composed of incredibly tiny strings, with average size of about a millionth of a billionth of a billionth of a billionth of a centimetre (called the *Planck length*). But this has just pushed the question of "What is a particle?" further back. We are now left with the question "What is a string?". We seem no nearer our goal in determining the ultimate bedrock foundation of reality.

In fact, this raises a potential weakness in all theories which believe Nature's secrets can be fully unlocked simply by examining the structure of the universe at increasingly smaller scales. The problem is: surely this "drilling" could just go on to infinity? How could it ever end? How would we ever know when we had reached the bedrock foundation of reality? As string theorist Brian Greene says in *The Elegant Universe*: "Another possibility, should strings fail to be the final theory, is that they are one more layer in the cosmic onion." Whatever structure we end up with, surely we will always be able to ask the question: "What is it made of?"

There is a famous (undoubtedly apocryphal) story which circulates in physics circles. A lecturer was presenting a lecture on astronomy. At the end of the lecture, a little old lady got up and said: "What you have told us is rubbish. The world is really supported on the back of a giant turtle." The lecturer gave a superior smile before replying, "What is the turtle standing on?" "You're very clever, young man" said the old lady, "but it's turtles all the way down."

We are supposed to laugh at the little old lady, but actually the lady might well be correct: reality might well be composed of successive layers of structural "turtles". In which case, approaches based on analysing structure at ever-smaller scales have got serious problems.

String theory and loop quantum gravity are potential candidates for a so-called theory of everything. However, it is hard to see how a theory based on structure can ever be a theory of everything – there will always be the question "Why this structure rather than another structure?" String theory attempts to get round this problem by proposing a multitude of parallel universes (called a *multiverse*) with different string configurations dominating in each universe. But, at the end of the day, the question would always still arise: "Why strings?" Why should the string structure (or loops if you subscribe to the loop quantum gravity approach) be favoured over any other possible structure of the universe? I am not suggesting that either string theory or loop quantum gravity is wrong, I just do not see how either theory could ever represent the fundamental theory of reality.

The Nobel laureate Steven Weinberg has continued this rather pessimistic tone: "I have to admit that, even when physicists will have gone as far as they can go, when we have a final theory, we will not have a completely satisfying picture of the world, because we will still be left with the question 'why?' Why this theory, rather than some other theory?"

I do not share Weinberg's pessimism. Surely there must be an alternative to the inexorable "drilling"? In his book *Three Roads to Quantum Gravity*, Lee Smolin suggests a third approach to quantum gravity research. He describes how a handful of researchers are starting with general *principles* and are starting to build-up from those principles in order to logically deduce the laws of Nature. It is this third approach which is followed in this book.

In contrast to the "drilling down from structure" approach, we will be "building-up" from fundamental principles. In this way, we will be certain that we will encounter the fundamental bedrock of Nature – because we will be building-up from it. After all, you wouldn't build a skyscraper – or any edifice – from the top-down: you would always start with bedrock and build up.

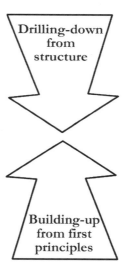

The principle of principles

The great advantage of theories based on principles is that they are not arbitrary (as opposed to theories based on structure). A fundamental principle (or *first principle*) appears obviously true. A fundamental principle contains the reason for its obvious correctness stated **within itself**. It is a self-contained entity, requiring no further explanation, validation, or support. A fundamental principle should appear so obvious that it would be impossible to conceive of a universe in which that principle is false. In this way, a principle is not arbitrary; instead it appears to represent a fundamental, unchanging truth which would have to be true in all conceivable universes. For this reason, a principle can form the very bedrock of the foundation of any true theory of everything.

This idea – that physics could be built up from deductive reasoning – has been considered for many centuries. In the seventeenth century, the philosopher Descartes proposed a system of physics based on reason. However, this methodology of logical deduction never gained a following, which I feel is a shame, as I believe the time is now right for it to play an important role in a physics in which experimental confirmation of theories is becoming ever-harder to achieve.

Our current particle accelerators can generate only a tiny fraction of the energy required to test theories of quantum gravity. The electroweak unification occurs with energies over 100 GeV (1 GeV is 1 followed by nine zeroes). The W and Z particles associated with electroweak unification have therefore been detected with our current accelerators which can reach energies of up to 14 TeV (1 TeV is 1 followed by 12 zeroes). However, the energies required to test quantum gravity will require energies of the order of the Planck scale of 1.22×10^{19} GeV (1×10^{19} is 1 followed by 19 zeroes). It might be many centuries before such an accelerator could be constructed, if ever. In the meantime, an alternative is the approach described in this book: an attempt to try to unify relativity and quantum mechanics by logical deduction from fundamental principles.

One of the great strengths of the approach of using general principles to derive theories is that there is no mention in principles of any particular structure or scale. Principles have such generality that they must apply to all physical processes in the universe, no matter if the underlying microscopic structure is strings or loops or any other arrangement. Also, as principles are completely invariant to scale, principles apply to the very small as much as the very large – they must apply to the realms of both quantum mechanics **and** relativity. For this reason, the method of building-up from principles is clearly ideally suited to the task of discovering a unification of relativity and quantum mechanics.

It must be admitted that there are pros and cons in the "building-up from principles" approach rather than the "drilling-down from structure" approach. The building-up approach means we have a solid bedrock in fundamentals, so we avoid the apparent arbitrariness of the drilling-down approach. However, it is that arbitrariness of the drilling-down approach that makes it so hard to duplicate. There are, for example, 19 numerical constants in the standard model of particle physics (known as *free parameters*), the values of which seem completely arbitrary. It is even conceivable that the universe could have been created with more or less than the three dimensions of physical space. It is hard to imagine how any approach which builds-up from obvious first principles is ever going to be able to capture that arbitrariness of structure. However, the "drilling-down" approaches seem to be having little success in this respect, either.

The question arises as to how much of the universe could have been created differently ("contingent"), and how much is a logical necessity. Our eventual goal is expressed clearly by Einstein: "What I am really interested in is whether God could have made the world in a different way; that is, whether the necessity of logical simplicity leaves any freedom at all." If a fundamental principle is true in every conceivable universe, then it would indeed possess a "logical simplicity" which would limit any freedom of design choices. By building-up from those fundamental principles we will make progress in discovering how many of the laws of Nature emerge as a result of logical necessity, and how many appear to require the assignment of arbitrary values.

Another reason to prefer theories based on principles is simply ... simplicity. At the ultimate base of Nature, we will surely find a principle which is beautiful and simple and obvious. The base of the tree would not only have to be simple, it would have to be "strong", for it would have to support the whole of the laws of Nature. The principle would have to be "strong" in the sense that it could clearly never be refuted (proved false). At the base, we would no longer expect to find a theory which could be derived from a more fundamental theory – because there could be no more fundamental theory. We would expect to find a principle, or, as John Wheeler suggested, an idea. But in order to reside at the base of the tree, it would also have to represent the end-of-the-line: a principle which was self-contained, a principle which required no further explanation, a principle which contained within itself the reason why it would be obviously true.

For our solution, we should be looking for something which is ultimately simple, and ultimately unified.

2

UNIVERSE

The universe.

The clue is in the title: "uni"-verse.

Uni-verse. Uni-fication.

The clue to unification is the universe itself.

"Uni" means "one". The universe is the one thing that exists. It is everything that exists.

In the time I have been working in physics, two themes have emerged to dominate my lines of thinking, two themes which appear to represent the underlying truth of our reality. Both of those themes appear to describe an aspect of the universe, revealing the structure of the universe to lie at the core of reality, and the properties of the universe being the key to obtaining ultimate unification.

Those two themes are both closely related to each other. The two themes provide two different ways of looking at the universe, but, together, they tell us the same thing. One theme is provided by quantum mechanics, and the other comes from relativity.

The common factor, the key to unification of those two theories, is the universe.

The clue always was in the title.

Connectedness

The first theme is "connectedness". Quantum mechanics tells us that there is no such thing as a completely isolated object: objects are connected. The impression of separateness is just an illusion.

It is possible for a particle to interact with another particle in such a way that the two particles form a single *entangled* quantum state. What this means is that the state of one particle is dependent on the state of the other particle in some way. Because of this dependency, it is a mistake to consider either particle in isolation from the other. Rather, we should combine the states and treat the result – both particles – as a single, entangled system.

For example, a light beam is composed of a stream of particles called photons. The direction of light's electric field is its direction of polarization. The polarization direction of a photon can be at any particular angle, for example vertical or horizontal. It is possible to generate a pair of entangled photons if, for example, a laser is shone at a crystal. In that case, a single photon can split to become two photons. Each photon produced in this way will always have a polarization orthogonal to the other photon. For example, if one photon has vertical polarization then the other photon must have horizontal polarization (this is due to the law of the conservation of angular momentum: angular momentum of the system before the split must equal the angular momentum of the system after the split).

So if two people each receive one of the entangled photons and each performs a measurement, they will find that the other person's photon has orthogonal polarization. This is not a big deal, you might think: there are two photons, and they have different polarizations. Whether they are separated by a great distance or not is irrelevant.

Well, in normal circumstances, you would be correct. But when we are dealing with quantum mechanics, objects can behave in quite counter-intuitive ways. As we shall see in Chapter Six, quantum mechanics tells us that before we take the polarization measurement of our photon, we have to consider the photon as having **both** polarizations, vertical **and** horizontal. Only after we measure the photon does it take a definitive value – either vertical **or** horizontal. So, if we are dealing with entangled photons, this means that the other entangled photon's polarization is only defined immediately after the measurement of its entangled partner. Bizarrely, this seems to indicate some instantaneous signalling through space, from one entangled photon to the other, saying "I've been measured – this is my value".

Einstein, in particular, was deeply unsatisfied with this result as *super-luminal* (i.e., faster-than-light) transmission of information is incompatible with the theory of special relativity. Special relativity states that the speed of light is a cosmic speed limit for transmission of information. However, it turned out that – in this case – Einstein appears to have been wrong, and the bizarre, counter-intuitive nature of quantum mechanics was correct. Einstein never accepted the implications of quantum mechanics. [2]

What this discussion of quantum entanglement really reveals, though, is that our human perception of objects being separated is not a match with the physical reality of the

───────────────────────

[2] Even though quantum entanglement appears to predict instantaneous communication, it does not, in fact, contradict special relativity as it is impossible to send information via entangled particles. For more about quantum entanglement, see my web page:
http://www.whatisreality.co.uk/reality_quantum_entanglement.asp

situation. Just because we can see no visual link between a pair of particles does not mean those particles are wholly separated. Indeed, quantum entanglement says we should not consider objects as being separated at all: we should treat all objects as one object. The message from quantum mechanics is that we should treat the universe as one connected, entangled object.

The one and only

The second theme is that the universe is the only thing that exists. There is nothing apart from the universe. If the last theme – connectedness – was associated with quantum mechanics, then this theme is associated with relativity.

As the theory of loop quantum gravity has its basis in relativity, we might expect the loop quantum gravity team to provide us with insight into the implications of the universe being the only thing that exists – and this is what we find.

In his excellent book *Three Roads to Quantum Gravity*, Lee Smolin introduced a simple principle: **there is nothing outside the universe**. If we define the universe to be the sum total of absolutely everything that exists, then there can clearly be nothing "outside" the universe. This implies that there can be no absolute axes of space or time outside the universe.

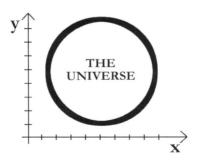

This, in turn, means that the position of all objects in the universe can only be defined **relatively** in terms of all other objects in the universe. Because of this relative definition, we would expect to see the laws of Nature behaving in a relativistic manner. And this is, of course, what we do see in the theory of relativity. So relativity is telling us that there is nothing outside the universe: the universe is the only thing that exists.

It is as if quantum mechanics and relativity are both telling us the same thing in different ways. Quantum mechanics tells us that everything is connected as one object: the universe. And relativity tells us that the universe is the only thing that exists. Two ways of saying the same thing. It is as if quantum mechanics is on the inside of the universe, looking out, telling us the universe is connected. Whereas relativity is outside the universe, looking in, telling us the universe is the only thing that exists:

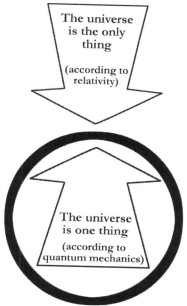

When the two great fundamental theories in physics are both telling you the same thing, albeit from different viewpoints, I think you should listen.

From these two different viewpoints of the structure of the universe, we obtain our definition of the universe:

u·ni·verse   (yoo'nə-vûrs')
n.

    1. The universe is one thing,
      the only thing that exists.

That single definition of the universe combines part of the behaviour of both quantum mechanics and relativity. In this respect, it represents a very basic form of unification in itself.

There are two parts to this definition: "The universe is one thing" stresses the message of quantum mechanics, whereas "The only thing that exists" stresses the message of relativity. The two parts essentially mean the same thing, but together we get the first hints of possible unification to come. The theme of the universe is one of unity.

From this definition of the universe we can derive Lee Smolin's principle "There is nothing outside the universe". Obviously, if the universe is the only thing that exists, then there can be nothing outside it. For this reason, this principle is obviously true. Moreover, this principle would have to be true in all conceivable universes – because there could be nothing outside **any** conceivable universe.

Hence we have at last found our fundamental principle, a principle which contains the reason for its obvious correctness stated **within itself**, a principle which is obviously true in all conceivable universes. So we have at last found John Wheeler's "utterly simple idea" which lies at the base of the tree. As John Wheeler said: "Oh, how beautiful. How could it have been otherwise?"

Lee Smolin referred to his principle as the "first principle of cosmology". We will see it is far, far more than that. Taking the role of our fundamental principle, we will discover that a remarkable proportion of the laws of Nature can be derived from this principle. The very fact that we can derive so many of the laws of nature by building-up from this principle will give us confidence that this is, indeed, the fundamental principle.

Absolutely no absolutes

Who, or what, is responsible for keeping the trains running on time? Fairly sensibly, you might suggest the train company is responsible, and this is surely the case. The train company has to maintain the tracks and the trains. It has to ensure the trains have enough fuel. It has to ensure the trains leave the station on time. Perhaps more fundamentally, it has to ensure the railway track extends the full distance between destinations!

But the train company is not the only entity which has responsibility for keeping the trains running on time: Nature has just as much responsibility. Nature has to ensure that its laws apply consistently to all objects. Nature has to ensure its laws do not drift over time. It is all well and good if the train company ensures its trains have enough fuel, but if Nature fails to ensure that the chemical reaction in the train engine produces enough energy, the train will not be going anywhere. Similarly, the train company can ensure its trains depart from the station at the correct time according to the station clock, but if Nature makes the clock run at a faster or slower rate then everyone will miss the train. Also, while it is essential that the train company ensures it has laid enough track to connect two destinations, if Nature allows any deviation in the length of the rails – such as a length contraction – the passengers are going to be stranded in the middle of nowhere.

Just as it is vital that Nature ensures the smooth-running of the railway, it is essential that Nature applies its laws in a consistent and unchanging manner to ensure the smooth-running of the entire universe. In this respect, Nature has a very responsible job, and it takes its responsibilities very seriously.

Now let us imagine Nature is expanding and diversifying its range of operations. It is going to use its profits from its highly-successful railway franchise and is going to move into the construction industry. Nature has bought a factory, and is going to supply a range of off-the-shelf, timber-framed "kit" houses which the customer puts together using only a screwdriver and some screws (such timber-framed kit houses do actually exist – the German HUF Haus company being the most famous).

In order to keep production costs down, the design is very simple: all the pieces of timber are one metre long. The construction of each house requires 1000 of these metre-long planks. So the job of Nature in this case, as factory boss, is to ensure that all the planks of wood produced by its factory are all precisely one metre long.

This raises some fundamental questions about the principle of measurement. For a start, it is vital to supply the *units* in which any measurement is made, e.g., metres, or kilograms. It is units which give our measurements meaning because units relate our measurements to some previously agreed-upon standard for some physical quantity.

For example, when we measure a plank of wood we are finding the relation between the wood and the marks on a measuring tape. Units provide the essential relational link between the unknown object being measured and the rest of the known world. And this will be a key theme in the later chapters of this book: giving property values to isolated objects is meaningless – property values arise through the interaction of an object with the rest of the universe. The length property of a plank of wood arises from its relationship with the measuring tape.

If we forget to give our measurements in units, or if we are not in agreement over which units we have both used to make our measurements, it is a recipe for disaster. For example, in 1999, the Mars Climate Orbiter space probe was intended to orbit Mars at a low altitude while mapping its surface. It was known that the probe could not get closer than 80 kilometres from the Martian surface or atmospheric stresses would rip it apart. However, the probe actually came within 57 kilometres of the surface and did, indeed, disintegrate. The crash investigators found that the cause of the error was due to the flight system software calculating thrust in metric units, while the ground crew were entering thruster data using imperial measures.

But probably the most notorious misuse (or absence) of units is from the 1984 rockumentary *This Is Spinal Tap*. In a famous sequence, lead guitarist Nigel Tufnel attempts to explain to a bemused reporter why his amplifier is louder than anyone else's because "Everyone else's amp only goes up to 10. This is 1 louder. This goes up to 11."

Let us return to consider the kit house company. If you remember, Nature – as company boss – has the task of producing 1,000 metre-long planks of wood. In order to do this, Nature has to measure the wood using a metre rule. Essentially, it is relating the wood to the known length of a single physical object (which, in the case of the metre, used to be the length of a platinum bar stored safely in a vault in France). This then becomes a standard of length on which everyone can agree. Everyone can relate their own measurements to this one physical standard stored in Paris.

The length of the platinum bar acts as an **absolute** standard: it has an independent existence, an independent reality. It is not dependent in any way on the people taking the measurement, or the objects being measured. It is as if the absolute standard has to exist entirely **outside** of the system which contains the object being measured, and the object performing the measurement. The platinum bar sealed in the vault in Paris is a good example of an absolute standard because it exists independently, entirely outside of the kit house factory.

Now try to imagine that the kit house factory is the entire universe, and it is completely under the control of Nature. You will be able to see why the seemingly innocuous statement "there is nothing outside the universe" has such tremendous implications. Nature can no longer access anything outside the universe – anything outside the factory. That means it can no longer access the absolute standard of the platinum bar in Paris. So the principle that "there is nothing outside the universe" reveals that all fundamental behaviour must take place in the absence of any absolute reference.

But, if you remember back to when Nature was in charge of running the railways, Nature takes its job very seriously and always does the best it can to keep the universe running smoothly. Even though Nature has lost access to its metre rule, **Nature always does the best it can with the tools**

available. If Nature cannot use an absolute standard external to the universe, it can still use a standard of length which has to be defined in terms of objects **inside** the universe. This means if it wants to measure objects inside the universe, it can only make **relative** measurements, relative to other objects in the universe – the option of taking absolute measurements is not open to Nature.

What does this mean for Nature running its kit house factory? Remember, Nature has the job of producing 1,000 metre-long planks of wood, and its job has just got considerably harder because it has lost access to its metre rule. But Nature is going to do the best with the tools available: it is going to take relative measurements, relative to other objects in the "universe". To achieve this, Nature picks one plank at random and declares this to be its new standard length (perhaps humorously called the *plank length?*), and it then compares the other planks to this standard plank and proceeds to cut 999 planks to exactly the same size as this standard length. Of course, this means that the planks are no longer guaranteed to be one metre long, but at least all the planks will be the same size.

But, here is the really strange thing: even though the planks are not necessarily one metre long … it doesn't matter! As long as all the planks are the same size, it is perfectly possible to construct the house. All the floors and walls and ceilings will fit together perfectly. The door will fit in the doorway, and the roof will fit with no problem.

Though, of course, there will be a side effect …

Side effects

So we have just seen that Nature – even though it has been denied access to an absolute standard – has managed to produce its kit house by cutting 1,000 planks to the same standard length. This has necessarily been a relative measurement: each plank has been measured relative to another randomly-selected standard plank to ensure all 1,000 planks are the same size.

However, when the customer buys the kit, and constructs it using a screwdriver, he gets something of a shock. While it is true that his house fits together perfectly – all the floors and walls and ceilings come together perfectly – he finds his finished house is twice the size of his neighbour's house – even though they bought the same type of kit house!

The reason for this discrepancy is fairly obvious: Nature was forced to use relative measurements – it was the only option open to Nature. While this resulted in a house which fitted together perfectly well, the size of the house could turn out to be any random size relative to other houses, depending on the random plank length which Nature picked as its standard length.

The final house is effectively selected at random from a virtually infinite range of sizes!

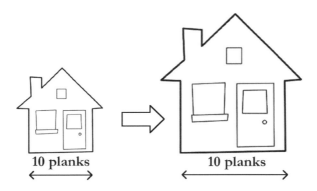

So being forced to use a relative system of measurement rather than having access to absolute standards is not a serious problem for the laws of Nature – systems can function perfectly well on that basis. But using a relative system of measurement can introduce unexpected, counter-intuitive behaviour which can appear weird.

So what kind of counter-intuitive behaviour do we experience in Nature, due to this lack of absolutes? First, let us consider the counter-intuitive behaviour due to relativity.

Just as there are no absolute axes of position outside the universe, there can also be no absolute axes of time: there is no absolute time in the universe. Just as the only meaningful measure of velocity in the universe is relative velocity, so the only meaningful measure of time is relative time. For example, it would only be meaningful to say "Your clock (and, hence, your measure of time) is running twice as fast **relative** to my clock." When Einstein published his special theory of relativity he did indeed show that a moving clock (and, hence, time itself) would run slow relative to the clock of an observer. It has been shown experimentally that an atomic clock in a jumbo jet travelling around the world shows less elapsed time than an identical clock which remains at base. This so-called *time dilation* predicted by special relativity represents a side-effect caused by Nature being unable to access an absolute time reference. We will be considering time dilation in detail in Chapter Four.

Another example of a side effect which is predicted by Einstein's theory of general relativity will be considered in Chapter Three. A speculative theory will be presented in which it is proposed that the gravitational field has to be defined relative to the masses in the universe. As a result, space itself can be curved around large masses. Normally, Nature does its job very well and it appears we live in flat space. But we will see a glitch if we study the path of light around a large mass, such as a star, we will see the light being bent around the star. We will see that this can be considered to be yet another side effect of Nature being unable to access absolutes.

Here we have seen that relativity has provided us with a number of examples of counter-intuitive behaviour. The reason why this relative behaviour appears weird to us is because we generally only deal with absolute measurements in our daily lives. When we go to buy clothes, we already know our own measurements according to a system of absolute units, and we know the clothes in the shop will also conform to that absolute system of units. Hence, it is quite easy to find clothes which fit. This is only possible because the shop, the customers, and the clothes could be considered as forming a *closed system*, and the absolute standard measure (that metre of platinum in Paris) resides outside of that closed system.

At human scales, we have access to absolute measurements. We can access measuring tapes which provide us with absolute measurements of length, and speedometers which provide us with absolute measurements of speed. However, when we get to the extremes, when we get down to the most fundamental levels, Nature is pushed to the limit and all of the absolutes get progressively stripped away. That is when we start to see the inevitable little "glitches" that are caused by the switch to relative measures.

So, we encounter fundamental behaviour in Nature which appears very weird and counter-intuitive to our eyes, but, really, it is only Nature behaving in a completely rational way, trying to make the best of the fact that – in all of its fundamental operations – it has to operate without having access to absolutes. We view this fundamental behaviour as weird, and try to find an underlying cause. But no underlying cause can ever be found, because this is fundamental behaviour. Nature's counter-intuitive fundamental behaviour is completely natural and is to be expected – the only problem is our preconception of what should constitute "normal" behaviour. In many ways, the relative behaviour of Nature should be considered to be "normal", and it is our absolutist view of behaviour which is a peculiar, special case.

We could call these bizarre effects **side effects**. Nature tries to do the best it can using the limited tools at its disposal (i.e., having to use relative measurements). For the most part, Nature does a very good job and it still appears it has access to absolutes. However, the side effects start to show through at the extremes: at extremes of speed, in the presence of extremely large masses, or at extremely small scales.

So here is an intriguing thought: another source of peculiar, counter-intuitive behaviour is seen in quantum mechanics at the very smallest scales. If quantum mechanics is a fundamental theory like relativity, might it not be possible that the root principle behind quantum mechanics is also the lack of absolutes in Nature? Might it not be possible that the counter-intuitive little glitches of quantum mechanics are also side effects, similar to the side effects seen in relativity?

This is a remarkable thought: could there really be such a simple link between relativity and quantum mechanics? In Chapter Six, which is dedicated to quantum mechanics, we will consider the counter-intuitive behaviour of quantum mechanics in detail, including the description of a famous experiment in which particles can apparently be in more than one place at once. But all the weird manifestations of quantum mechanical measurements can be described as being variations of a particular characteristic quantum behaviour: before measurement of a quantum property, we find the property appears multi-valued, and when we perform the measurement of the property we find we get a random value from a potentially infinite list of possibilities. **But this is exactly what we found with our kit house when we sold it to a customer!** When the customer has constructed his house, he finds his final house could be any size from a potentially infinite range of sizes, and the selection of the type of house he receives is random (depending on the length of the randomly-selected plank in the factory). **This has clear parallels with quantum mechanical behaviour!**

So the lack of absolutes in Nature means that Nature is forced to switch to a relative mode of operation. It is known that this introduces counter-intuitive side effects in relativity, but the hypothesis presented in this book proposes that this also introduces bizarre side effects which are interpreted as quantum mechanical behaviour.

The basis of the hypothesis presented in this book is that the counter-intuitive behaviour of both relativity and quantum mechanics has its roots in the lack of absolutes in Nature: two radically different theories, with one common source.

I said earlier that the behaviour of both quantum mechanics and relativity is telling us about the unified structure of the universe – and we should listen. But, on the basis of this discussion, we now realise this should be the other way round: **it is the unified structure of the universe which is telling quantum mechanics and relativity how to behave.**

Parallel universes

If we return to consider our principle that there is "nothing outside the universe", the question might be arising in the minds of some readers: "What about parallel universes? Surely, if such a thing existed, it would be outside our universe? In which case our presumed principle – that there is nothing outside the universe – would be wrong." This requires investigation.

Parallel universes are very much flavour-of-the-month in physics at the moment. The concept appears to offer an easy solution to some of the greatest current challenges in physics. It could be argued that this is not necessarily a good thing, and I would tend to agree.

At the current moment in time, fundamental physics has found itself in something of an impasse. Progress is slow, and there are several deep and troubling mysteries – with no obvious solution – on the horizon. These questions include the unknown composition of dark energy which is apparently responsible for the observed acceleration of the expansion of the universe. Another mystery concerns the apparent fine-tuning of some of the fundamental constants. It has recently been realised that if some of the fundamental physical constants of the universe were only slightly different then the existence of life in this universe would have been impossible. The most notable example of this is the apparent fine-tuning of the *cosmological constant* which governs the expansion of the universe. It appears the cosmological constant has a tiny value, but is not zero. It the constant was much larger, the additional acceleration of the universe would have prevented stars and galaxies from forming. And if the value was much smaller, the universe would have collapsed back in on itself before life could have developed.

The principle of parallel universes offers a simple way out of this conundrum: just postulate the existence of a range of different universes, and a random mechanism by which the fundamental constants vary from universe-to-universe. Then it is possible to present a simple explanation as to why the constants are set to such fortuitous values in our universe: there is nothing special about our universe, we just happen to reside in a random universe in which the constants are set correctly.

However, the weakness of this so-called *anthropic* reasoning is that it can be used to predict anything, and a theory which predicts anything predicts nothing. It seems as though we are giving up finding a theory which unambiguously predicts the nature of the universe. As Brian Greene says in his book *The Elegant Universe*, this method has the capacity "to lessen our insistence on explaining why our universe appears as it does."

I am definitely not a fan of parallel universes, but we must consider the potential implications of them representing something "outside the universe". To answer this question, we will refer to a comprehensive review of parallel universes conducted by Max Tegmark. [3]

Theoretical parallel universes actually come in a variety of different flavours. The first type of parallel universe considered by Max Tegmark arises from the theory that space is infinite (this is by no means certain – space might be *closed* in which case it has a finite volume instead of being infinite). The theory states that if space truly is infinite, and uniformly filled with matter, then somewhere out there in the dark depths of space there is going to exist every possible structure. Hence, somewhere far, far away, beyond the limits of our visible universe, there exists a pink hippopotamus playing the trombone to Marilyn Monroe. It sounds like a joke, but this is actually a prediction of the theory. As Max Tegmark says: "Even the most unlikely events must take place somewhere."

---

[3] *Parallel Universes* by Max Tegmark, http://arxiv.org/abs/astro-ph/0302131

As you can see, this theory makes some rather bold assumptions about the rather dubious physical quantity of "infinity". But, more importantly for this discussion, does this represent a separate "parallel universe" in any sense?

This type of parallel universe is identified purely on the basis of its distance from our visible horizon. Basically, it is a long, long way away – that is the only distinguishing feature of this region of space. But Antarctica is a fairly long way away as well. Does this mean Antarctica is in a parallel universe? If not, then where do we draw the line? How far away has the region of space got to be before it is denoted as being a parallel universe?

Surely, this type of "parallel universe" has as much claim to being a parallel universe as Antarctica does: in other words, none whatsoever. A distant region of space is still part of our universe.

The next type of parallel universe we will consider is perhaps more in tune with the general conception of what a parallel universe should resemble – mainly through their rather clichéd use in popular science fiction. This concept of a parallel universe is of a completely separate realm of existence, distinguished by having its own completely distinct spacetime structure to that of our universe. This is the type of parallel universe much beloved of many a Star Trek episode, for example, the episode in which a transporter mishap swaps Captain Kirk and his companions for their evil counterparts from a parallel universe – the episode entitled *Mirror, Mirror* (I'm giving my age away, here).

Such a universe could theoretically exist on the other side of a black hole, the black hole providing a *wormhole* to another universe. Another possible source of this type of separate realm of existence is provided by the *Many Worlds* interpretation of quantum mechanics (which we will consider in detail in Chapter Six).

But, a universe which exists as a separate realm of existence, which can have no effect on the objects in our universe, and which itself can never be observed, surely does not exist – by definition. As Paul Davies says in his book *The Mind of God*: "There is something philosophically unsatisfactory about all those universes that go unobserved. To paraphrase Penrose, what does it mean to say that something exists that can never in principle be observed?"

However, to be honest, these distinctions are all irrelevant for our current discussion in this book. This is because we are completely at liberty to define the term "universe" as we wish, and we are choosing to define the term in the loosest possible manner as "the sum total of absolutely everything that exists". So if we decide that parallel universes exist – even though we can never observe or interact with them – then they are, by definition, part of the "universe".

The universe is defined as everything that exists. Whether or not this means the universe is composed of a subset of so-called "parallel universes" is irrelevant.

The universe is everything that exists. Therefore, there is nothing outside the universe.[4]

---

[4] The principle "there is nothing outside the universe" should not be taken as making any profound statement about the existence of God. After all, a God which exists would be a part of the universe – by our definition, the universe being defined as "everything that exists". This is a book about quantum mechanics and relativity – other questions about deeper philosophical and theological issues are beyond the scope of this discussion and can be left to other, more contentious, books.

Hypotheses about the nature of reality which invoke parallel universes are very popular at the moment. These hypotheses certainly seem to infest popular science books! However, I would consider such hypotheses – which can, by definition, never be falsified – to be borderline unscientific. Not only that, but I would consider these hypotheses to be positively dangerous for science, for two reasons.

Firstly, as Brian Greene explained, invoking parallel universes makes the job of physics too easy. Instead of seeking theories which explain how the universe works, theories which make testable predictions, we end up with hypotheses which explain nothing. Absolutely anything can be explained by invoking anthropic reasoning: just say the laws of physics are different in each parallel universe, and we just happen to inhabit a universe in which the laws are set to those which we observe.

But, secondly, I feel parallel universe ideas are dangerous because they are a distraction, a diversion, from focussing all our attention on the one universe we inhabit. The premise of this book is that there is one universe, and that is all there is. The universe is a strange, beautiful, and surprising construct, and all the solutions we seek are contained within this universe – if we just direct our gaze at it instead of being distracted by fanciful hypotheses.

If this book forms just a tiny part of a fightback against parallel universe hypotheses then I would be delighted. As far as this book is concerned, our universe is the one and only.

3

SPACE IS NOT A BOX

The discussion so far has focussed on the lack of absolutes in the universe. So now let us examine the implications of this lack of absolutes in space and time. This chapter is devoted to considering the implications of the lack of absolutes in space. The following two chapters will be devoted to considering its effect on our concept of time.

Let us start our analysis of how our everyday concept of space must be altered.

As soon as we ask about the nature of space we very quickly find it to be something about which we all have an intuitive notion, but which is very hard to define. And, unfortunately, space is one of those concepts about which human beings have very strongly-held preconceptions which are very hard to shake. Space is a concept with which we are all intimately acquainted, which we have known from birth. We all deal with space as we appear to move around "in" it all day long. Most people like space. They pay lots of money to get more of it, be it a bigger apartment or a bigger garden.

In this respect, space feels like a "real thing" – it costs money to buy. This is the picture we have of space. We know how it works, and we resent being told by logic or experiment that our personal, deeply-held concepts are wrong. This is a particular problem for space as it does, indeed, appear that our preconceived notions are very far from an accurate model of the reality. It turns out that space is not a "real thing" after all, so if you have spent a lot of money buying some recently, then maybe you should take it back to the shop!

Isaac Newton believed in the concept of absolute space, which meant it was possible to define the position of each point in three-dimensional space. Even if all the matter in the entire physical universe was removed, Newton believed absolute space would still exist as an entity in its own right – as if space was a three-dimensional box, and the physical universe was positioned at some point in the box. The position of any object could then be absolutely defined by some sort of measurement scales along the sides of the box:

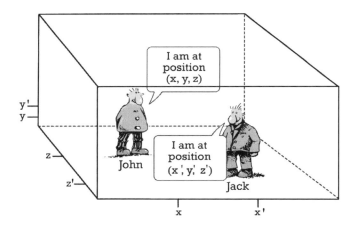

This absolute view of space is perhaps the most intuitive picture of space. It feels very natural for us to imagine space in this way, and there was perhaps an evolutionary advantage in developing spatial awareness which modelled space along these lines.

So now, consider a little story. Recently, I decided I needed more room in my apartment, so this morning I made a start putting all my old books and CDs into storage. I bought four large transparent boxes, and I put most of my books into those boxes and slid them under my bed. It's a good solution – boxes are really good for storage. We have all lived our lives knowing that we can always put things into boxes for storage.

Now let us imagine Nature has seen us putting things into boxes and so Nature decides it wants to put things into boxes as well. Specifically, Nature decides it wants to put the universe into a box. But how can Nature find a box into which it can put the universe? The universe – by definition – is everything that exists. Therefore, all boxes must be **inside** the universe; there can be no boxes outside the universe. So Nature is stuck. Nature cannot find a box into which it can put the universe into storage. There are no boxes outside the universe. Nature will just have to move to a bigger apartment.

As we have just described, Newton believed absolute space would exist as a box in its own right, even if all the matter in the universe was removed. However, if there is nothing outside the universe then how can there possibly be a box external to the universe? There can surely be no system of coordinate axes outside the universe we can use to determine the absolute position of an object.

With this in mind, we find Newton had a competitor with different ideas, the great German mathematician and philosopher Gottfried Leibniz. The story of the competitiveness between Leibniz and Newton is fascinating. It certainly was not a friendly rivalry, with Newton giving an impression of being rather ego-centric. It is believed Leibniz discovered mathematical calculus a few years before Newton, though Newton disputed this, and much of Europe found it hard to believe that Leibniz could have discovered calculus independently of Newton. By the time of his death,

Leibniz's reputation was in decline (Newton played a rather dubious role in this decline). However, Leibniz's reputation has been recently restored, and he is perhaps now held in higher esteem than was ever the case when he was alive.

Leibniz rejected Newton's idea of absolute space. Leibniz believed that space was defined only on a relative basis, with the position of objects being defined in terms of the position of other objects, not defined via some absolute coordinate system. If we state the distance between objects, then we define a geometry. We then get an impression of space emerging from those relationships, without space existing as an independent entity in its own right. If we removed the objects and their relationships, then space would not exist on its own. Due to this elimination of space as an entity in its own right, philosophers call this theory *eliminative relationalism*.

Leibniz had a famous debate via a series of letters to Samuel Clarke, an English supporter of Isaac Newton, between 1715 and 1716. Leibniz asked Clarke the hypothetical question: if the whole universe was moved ten feet to the left, how would we ever know? According to Newton, the two scenarios were distinctly different (positioned differently in absolute space), but for Leibniz, all the spatial relationships were maintained so the two situations were equivalent. Clarke found it hard to argue his position against Leibniz. Leibniz's argument for relative space was compelling.

The diagram opposite shows Leibniz's model of relative space – with space emerging as a result of the relationships between the men. Contrast it with the diagram of absolute space ("two men in a box") earlier in the chapter. Instead of each man having to store an (x, y, z) position property as was the case in absolute space, each man now has to store the distance between himself and each of his fellow men. This representation still maintains all of the information which was present in the absolute space example:

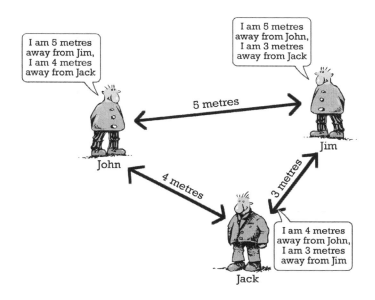

Do not be fooled by the diagram above of the three men in relative space appearing to be separated by some form of "distance" in "space" – this is just a weakness in the way the diagram is drawn. The men are not "in" space (as was the case in the earlier diagram showing them in absolute space). The arrows merely show the relationships between the men – they should not be considered as physical arrows with an extent in space. Remember, space does not exist as an entity in its own right: if all the matter in the universe was removed, there would be nothing left – no "space" left with an independent existence of its own. As Einstein said: "Spacetime is not necessarily something to which one can ascribe a separate existence, independently of the actual objects of physical reality. Physical objects are not *in space*, but these objects are spatially extended. In this way, the concept *empty space* loses its meaning."[5]

If there is nothing outside the universe – and, logically, there cannot be – then absolute space is clearly refuted: Leibniz was surely correct. We do, indeed, live in Leibniz's relative space, not Newton's absolute space. However, towards the end of the chapter we will see how Newton proposed a powerful counter-argument to Leibniz which led to Newton's idea dominating for two centuries.

Symmetry

We are probably all well–acquainted with the concept of symmetry, but perhaps have not given much thought to the question of why something is considered symmetrical. We are probably most familiar with geometrical symmetry, such as a snowflake:

If you were to rotate a snowflake by 60° then it would appear unchanged. This is because the snowflake is said to possess *rotational symmetry*. Similarly, if you were to view the snowflake in a mirror, it would look unchanged (*mirror symmetry*). So this gives us a better idea of how to define

---

[5] Albert Einstein, *Relativity*, 1916

symmetry: if something remains unchanged after a transformation, it is said to possess symmetry.

Any type of transformation could reveal symmetries, not just rotations and mirrorings. For example, we might consider the transformation of *spatial translation*, in other words, simply moving an object to a different location. If an object is defined relative to an absolute axis incorporating a scale, then it is not possible to translate an object without changing the physical situation. This is how Newton's absolute space would function. Considering the image below (showing absolute space), if we were to move the ball from position 5 to position 7, then the physical situation would have altered from "the ball is at absolute position 5" to "the ball is at absolute position 7":

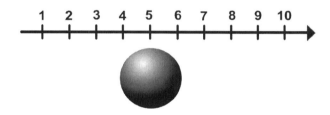

However, if the object is not defined relative to an absolute axis with a scale, then the object can be translated at will, and the physical situation will not be altered:

This is how relative space behaves.

So, the nature of relative space ensures that all objects possess spatial translation symmetry. Fairly obviously, if you were to perform an experiment, and then perform the same experiment a second time but this time with the apparatus moved six foot to the left, you would get the same result. The experiment would be unchanged after the transformation of spatial translation. In fact, everything in the universe has spatial translation invariance. This is due to the lack of any absolute scale "outside the universe" to differentiate any particular position from any other position.

So perhaps we could propose a deeper explanation for symmetry in physics: **symmetry reflects Nature's fundamental inability to distinguish between one physical situation and another**. If Nature is unable to distinguish between two situations, then we can transform one physical arrangement into another without altering the situation. This represents a symmetry.

As has been said earlier in this book, Nature always does the best it can with the tools available to it. However, Nature is limited by the lack of absolute scales available to it, which is why we see so many symmetries in the physical world. The concept of symmetry is extremely important in physics, and it will be a recurring theme throughout this book. Just remember that whenever you hear symmetry mentioned it is a result of Nature's inability to distinguish between physical situations.

Now let us consider another implication of symmetry on the laws of Nature. Imagine you are an astronaut, floating in the darkness of space. You are so far away from any planets or stars that all you can see is darkness. You feel completely stationary. Suddenly, on your right-hand side, you see a fellow astronaut floating towards you at a constant velocity. Your friend passes by you quite closely, before vanishing to your left-hand side into the darkness of space.

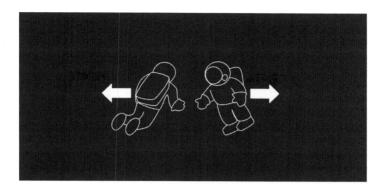

This story can be described from the point of view of the other astronaut, who will describe exactly the same experience as you – a symmetrical experience. The other astronaut felt completely stationary in space, he saw you appearing on his right-hand side floating by at a constant velocity, and it was you who drifted off into the darkness of space. So, the question arises: who was stationary and who was moving? Which of the two astronauts has the greater claim to be stationary?

In the hypothetical situation of absolute space, Nature could endow every particle with an (x, y, z) position property (just as it endows every particle with a momentum property, for example). Nature could then easily determine which astronaut was stationary and which was moving. Nature could tell that the stationary astronaut's position was unchanged, whereas the moving astronaut's position would be altered. As we discussed in the last chapter, though, Nature can only do the best it can with the tools available to it. Without access to absolutes of space, there is no way for Nature to distinguish if one astronaut is in uniform motion or if one is stationary. Hence, by definition, they are fundamentally in the same state of motion.

And if Nature cannot tell the two situations apart – uniform motion and stationary – then you will naturally feel the same experience in both situations. If you were sitting in an enclosed room in an office block, you would consider yourself to be stationary, but if you were sitting in an enclosed room in an ocean liner moving smoothly on a calm sea you might also consider yourself to be stationary – you would certainly feel as if you were stationary.

(Note the emphasis on uniform motion, i.e., motion not undergoing acceleration. If acceleration is involved, the situation is profoundly different as Nature can tell the difference – as we will see at the end of this chapter).

This inability of Nature to distinguish types of motion forms the basis of Galileo's unification of "moving" and "being stationary" known as *Galilean invariance*.

So here we have an example of a unification – the very first historical example of a unification – which is caused by Nature being unable to access absolute values. As discussed in the previous chapter, it is the belief of this book that the most pressing question in current physics – the unification of relativity and quantum mechanics – can also be achieved by realising that the two phenomena are both caused by Nature being unable to access absolutes. There might be a wonderful symmetry here, if the general principle behind Galileo's first unification provides the solution for a modern unification.

The gravitational field

But how is this relative space defined? What determines the structure of space? What controls the motion of objects? This is achieved by an entity which permeates the universe: the *gravitational field*.

So what, exactly, is a field?

If we have an electric charge, either negative of positive, and we move it through a region of space, we will find it either attracts or repels any other electric charges around it, even without directly touching them. It appears the first charge can directly move a second charge, even if there is nothing in-between the charges, and the two charges are separated by a distance. It appears as if the first charge has an "invisible hand" some distance away from it which could instantaneously move the second charge. This principle is called *action-at-a-distance*.

A similar situation applies to the force of gravity. Newton realised that the force which pulled an apple from the tree was the same force which held the Earth in orbit around the Sun. Newton explained this force of gravity in terms of action-at-a-distance, as though the Sun extended a giant hand to hold the Earth in its orbit. But how can something move something else across the vacuum of empty space without touching it, or apparently directly influencing it in any way? Newton was certainly not happy with his own conclusion:

> *That gravity should be innate, inherent and essential to matter, so that one body may act upon another at a distance thro' a vacuum, without the mediation of anything else, by and through which their action and force may be conveyed from one to another, is to me so great an absurdity, that I believe no man who has in philosophical matters a competent faculty of thinking, can ever fall into it. Gravity must be caused by an agent acting constantly according to certain laws; but whether this agent be material or immaterial, I have left to the consideration of my readers.[6]*

---

[6] *Essays* (1625), quoted in *The Oxford Dictionary of Phrase, Saying, and Quotation*, p. 34

Indeed, Newton's instincts were correct: action-at-a-distance does not appear to be an accurate picture of reality. Considering our electric charges again, we find any movement of the first charge is not instantly reflected by movement of the second charge (instantaneous communication being forbidden by special relativity). Instead, the influence of the first charge spreads out through space in all directions at the speed of light, taking time before it reaches the second charge. So, instead of action-at-a-distance, the accurate physical picture is of the first charge being the source of a *field* which extends out into space. The field is a continuous structure, which is defined at every point in space. When the first charge moves, it causes a disturbance in the electric field, and this spreads out through the field – as if you had dropped a rock into the sea, and the waves spread out over the water. Only when the disturbance reaches the other particles do they move accordingly (repelled or attracted).

In the case of the electric field described here, when we move a charged particle it creates a changing electric field. In turn, that changing electric field generates a magnetic field. That magnetic field generates an electric field, and so on in an oscillatory fashion. The resultant self-sustaining wave is an *electromagnetic wave*, and can carry energy across empty space. Visible light is a form of an electromagnetic wave, so we can certainly see types of fields: fields have a reality, they are not invisible figments of our imagination.

So a field is a structure which spreads through space, and when we put a particle in it, the particle feels a force. Hence, we could view gravity as a field – the gravitational field – which extends throughout all space and exerts a force on all particles. According to Newton's action-at-a-distance model, if the Sun disappeared then the Earth would instantaneously fly off into deep space, no longer held in place by the Sun's invisible hand. However, Einstein realised that such instantaneous action-at-a-distance was incompatible with special relativity which prohibited faster-than-light transfer of information. Instead, Einstein revealed that gravity should be explained in terms of a gravitational field, and the motion of the Earth around the Sun can be explained by the nature of the field in the immediate vicinity of the Earth. If the Sun disappeared, the Earth would continue to orbit the Sun's position for seven minutes, until the bad news reached the Earth via the gravitational field.

Einstein's insight occurred when he realised that an observer in a closed box undergoing constant acceleration would not be able to distinguish the force he was feeling from the force of gravity. Conversely, an observer falling freely in a closed box would experience no force of gravity. If a force can disappear purely because the point of view of the observer changes, then this means the force does not exist: it is dependent on the observer, it has no objective reality (i.e., not all observers experience it as real). It is not a real force. Therefore, Einstein realised gravity was just a fictitious force.

Not all forces are fictitious, and disappear with a change in the observer's perspective. As a comparison with the electromagnetic force, if an observer in a closed box has a particle with a positive electric charge, we could repel that positive particle using a positively-charged particle outside the box. However, we could then move the observer's box away in the direction of the repulsion to make it appear as though no force was being applied. In this way, it would appear we could make the observer think that no electromagnetic force was being applied to his particle. So, does that mean the electromagnetic force is a fictitious force like gravity? No, because if the observer had a negatively charged particle as well as a positive charge, he would find the negative charge attracted to the external positive charge with increased force. So it is not possible to make the electromagnetic force disappear merely by moving the observer: the electromagnetic force is a real force, not a fictitious force like gravity. We can only deduce this about gravity because gravity has the same attractive force on all objects – unlike the electromagnetic force.

So if gravity is a fictitious force, what is the cause of the force we all feel due to gravity? To answer this, consider an object moving freely in space, without undergoing any acceleration. Such an object would tend to continue in a straight line in space (called a *geodesic*) and would not experience any force due to gravity. Only if a wall, or any solid object, was placed in its way to restrict its movement would the object feel the force we recognise as gravity. The observer would feel the solid object pushing back against him, restricting his progress. This is just how we feel when we are standing on the surface of the Earth. Our natural motion should be towards the centre of the Earth, but the planet's surface restricts our motion. The resultant force we feel, making us tired when we stand up on the surface, is interpreted as the force of gravity.

What determines the nature of the force which we interpret as gravity? Clearly, it is anything which modifies our straight line through space – anything which modifies the geometry of space. In other words, gravity **is** the curvature of space. As we continue in what feels to us like a straight line in space, if space is curved then our motion will be deflected. It is this deflection of motion which is interpreted as gravitational pull.

What could curve space? Well, we know that objects appear to be attracted to large masses, such as planets, by the force of gravity. This means that large masses must curve space around them so that an object, while proceeding in a straight line in curved space, is deflected toward the mass. A two-dimensional demonstration of the principle behind the curving of three-dimensional space involves placing a heavy ball in the middle of a rubber sheet. The weight of the ball deforms the sheet downwards. The heavy ball is supposed to represent a large mass, for example, the Sun. If we roll another smaller ball around the sheet, it will roll in circles around the indentation. If we consider the smaller ball to represent the Earth, we can see that its resultant motion effectively orbits the Sun.

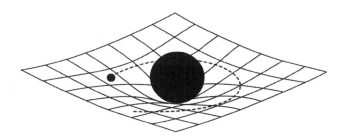

After observing these orbits, you might come to the same conclusion as Newton that there is an attractive force between the objects, holding the Earth in an orbit. Whereas, all that is happening is the Earth is following a straight line in curved space. As John Wheeler famously and succinctly explained: "Matter tells space how to curve, and space tells matter how to move."

In summary, large masses curve space around themselves. Objects moving freely, without being acted on by forces, travelling in a straight line in space appear drawn to the large masses purely due to the curvature of space. We interpret this deviation of motion as being due to a "force" of gravity, but really the force is completely fictitious – the object is merely trying to proceed in a straight line.

In the absence of any "box" of absolute space, it is the gravitational field which determines the way objects move, determines the distance between objects, and hence defines the geometry of space. Hence, the gravitational field plays a role which is identical to that played by space. We can consider the gravitational field to be synonymous with space: the gravitational field **is** the structure which we interpret as space.

Why is there a gravitational field?

The question might be asked, why is there a need for this peculiar invisible entity which pervades the universe, an entity which is curved by masses in a peculiar fashion, and which controls the motion of planets. It is worth investigating this question because, again, it appears that the answer to this question is that Nature is unable to access any absolutes outside the universe.

If Nature operated in absolute space, it could endow every particle with an (x, y, z) position property. This would make it very easy for Nature to track and control objects. However, absolute space is not an option – Nature has to work in relative space. Remember the golden rule: **Nature does the best it can with the tools available to it**.

In relative space, the positions of the objects could be determined by storing the distances between objects – just as in the earlier diagram showing the three men standing in relative space, with arrows indicating the distances between the men. All the information about the spatial arrangement of the men is retained in this representation. Nature could store the distances between objects as properties of those objects.

However, what happens when our universe starts to expand to include more than just these three men? For every new object introduced into our universe, the men would need to store a new property: the distance to the new object. For a real universe, containing all its objects, the list of properties which would have to be stored by each object would soon become virtually infinite in length!

A more practical solution would be for Nature to introduce a series of separate "relationship objects", separate from the men, the function of each of these relationship objects merely being to store the distance between two objects. Each new spatial relationship, for example, between a man and a tree, would be represented by an additional relationship object. In this way the "man" object no longer needs to hold a virtually infinite list of property values.

However, this structure of our relationship objects would be hugely redundant: it would have to hold the distance between every object in the universe and every other object in the universe. But, there is no need to hold such a vast amount of information when so much of the information would be redundantly duplicated. All that is needed is to find the distance between each point in space and the point infinitesimally close to it. For objects separated by distance, all these small differences could be summed. Clearly, this entity which spreads throughout the universe, which defines distances, and is defined at every single point forms a *metric field* (a field which defines measurements) and is a perfect analogy for the gravitational field! [7]

What seems to happen is that as soon as matter is introduced to the universe, Nature has to generate a gravitational field structure as the most efficient tool for describing the positions of objects (i.e., generating "space") and controlling the motions of objects. Note again that this would not be a problem for a universe composed of absolute space as, in that case, Nature could endow every particle with an (x, y, z) position property in the surrounding "box" of space – space existing as an entity in its own right. This is only a problem for the relative space situation, in which space has no independent existence. In relative space, Nature needs to generate a gravitational field.

---

[7] For a similar discussion, see:
http://plato.stanford.edu/entries/spacetime-holearg

Building-up from first principles is bringing us results. Our fundamental principle – that there is "nothing outside the universe" – has led us to the inevitable conclusion that we live in relative space, and that conclusion has led in turn to the prediction of the necessity of a gravitational field.

Background independence

So, as we have just discussed, the gravitational field must be a completely relational structure, with no dependence on any background coordinates (i.e., absolute space). Einstein realised this, and his quest for a so-called *background independent* theory of gravity took him eight challenging years. What background independence really means is that, rather than being defined in terms of coordinates in a fixed arena of absolute space, the shape of the gravitational field can only be defined by the distribution of masses in the universe. Hence, the final result is a fully relational structure: its shape is defined relative to the masses.

This principle of background independence is a guiding feature of the theory of loop quantum gravity. This is to be expected, as loop quantum gravity is based on general relativity, itself a background-independent theory. Loop quantum gravity theorists tend to emphasize the background independence of their theories for a very good reason: string theory is **not** background independent. In string theory, the strings are presented as moving against a fixed background of space and time.

I cannot help getting the feeling that the string theorists must be getting rather fed-up of the constant hectoring from the loop quantum gravity theorists about string theory's inferiority in this respect. However, from our analysis, we can see that the loop quantum gravity theorists have a good point: any theory which is not background independent (such as string theory) and effectively requires there being a "box" outside the universe is logically inconsistent (fatally flawed). Such a theory could never represent a truly fundamental theory of Nature without modification.

Absolutes strike back

Let's return to our discussion of absolute space earlier in this chapter. If you remember, Leibniz did not believe in the existence of absolute space. According to Leibniz, the positions of objects were only defined in relation to other objects. By extending this reasoning, we can see that – according to Leibniz – it should be possible to treat any object as stationary, with the rest of the universe moving around it.

However, Newton posed a counter-argument to Leibniz which effectively killed-off Leibniz's idea of relative space for two centuries. Newton noted that the surface of water spinning in a bucket would become concave, with the edges of the water rising up the sides of the bucket. Newton asked the question: relative to **what** is the water spinning? You might suggest the water is spinning relative to the environment and the objects surrounding the water, but if you consider the bucket (which is the water's most immediate environment) you will see that the water is spinning at exactly the same speed as the bucket. So the water is not moving at all relative to its immediate environment.

According to Leibniz, objects are only defined relative to other objects, and so it was just as valid to suggest the water was stationary and it was the rest of the world was spinning. This meant that Leibniz's argument could not explain the change in the shape of the surface of the water. If the water was stationary, why was its surface being deformed? Newton argued that if the water was not spinning relative to its immediate environment then it had to be spinning relative to absolute space, and its movement relative to absolute space was what was causing the deformation of the water surface. Leibniz was forced to admit defeat. Is our proposed concept of relative space in trouble?

However, in the mid-1800s, the Austrian physicist Ernst Mach presented an alternative to Newton's absolute space. Maybe the disturbance of the water in the bucket was not due to it spinning relative to absolute space, maybe it was because it was spinning relative to the rest of the world around it. Even though the water was not moving relative to the bucket, it was spinning relative to the ground, and the trees, and the sky. Wasn't this enough to determine that the water was spinning? Does Nature really need absolute space to tell it that the water is spinning when it has all these other clues?

In his superb book, *The Fabric of the Cosmos*, Brian Greene presents a discussion of Mach's principle which really provides food for thought. Imagine you are an astronaut floating in empty space, many millions of miles away from the nearest stars or planets. Now, imagine you are spinning in that position. Clearly you would feel like you were spinning because of the acceleration you would feel. Your arms would splay outwards, and your tummy might feel funny. Newton would say you experience the spinning feeling because you are spinning relative to absolute space. Mach disagreed with Newton's notion of absolute space, and sided with Leibniz's concept of relative space. Mach therefore argued that no viewpoint in the universe should be preferred. According to Mach, it is therefore just as valid to consider the astronaut is stationary and the rest of the universe as spinning. In that case, Mach argued, the only reason you feel you are spinning is because you can clearly see distant stars and planets around you, and so you are clearly spinning relative to the rest of the universe.

According to Mach, the rest of the universe forms a reference by which Nature decides whether or not you are spinning.

But here is a fascinating twist: consider you are spinning in space in which all other matter – planets and stars – has been removed. You can now see no other reference points to determine if you are spinning or not. And the same challenge applies to Nature. Remember, Nature can only do the best it can with the tools available to it. We know Nature cannot access any absolute axes of space, and now we have removed any other references by which it could judge you to be spinning or stationary. Nature appears to be fundamentally limited in what it can say about your state of rotation. For Nature, surely, it is fundamentally impossible to say if you are spinning or not. For this reason, surely you would not feel as if you are spinning?

It turns out that the solution to this paradox is that when you are spinning, you are actually accelerating: your velocity (speed + direction) is being constantly modified because your direction of motion is constantly being modified. But why should acceleration be able to make you feel as if you are moving, whereas movement at a constant velocity only makes you feel as if you are stationary? This is because Nature is able to distinguish accelerated motion from uniform motion. For this reason, science often describes accelerated motion as an "absolute" (note my quotes).

So, if acceleration is an "absolute", does this raise a problem for the premise of this book, i.e., that the universe contains absolutely no absolutes? We will be answering this important question in the next chapter.

Gravity: another side effect?

When Mach proposed his alternative to Newton's absolute space, he was attempting to provide an answer to the question of how acceleration can be defined in the absence of absolute space. Mach's answer was that accelerated motion could be determined relative to all the other masses in the universe – no absolute space was necessary. In proposing this solution, Mach realised that, in relative space, there is simply no alternative other than defining all position (and, therefore, motion) relative to all the other masses of the universe. There is no other option available to Nature.

Even though Einstein's concept of a gravitational field superseded Mach's idea, this basic principle of Mach still holds true: in relative space, there is simply no alternative to defining position in terms of the other masses in the universe. So this principle inevitably applies to the gravitational field. Being unable to access any absolutes outside the universe, Nature is forced to make the gravitational field an entirely relational structure, i.e., not being defined in terms of the coordinates of absolute space **outside** the universe, but having to be defined in terms of the structures which exist **within** the universe. In other words, the gravitational field has to be defined in terms of the distribution of matter inside the universe – Nature has no option. This dependency explains why, when matter moves in the universe, the gravitational field is inevitably re-shaped.

Although this is a speculative theory, we should not be surprised to find that this inevitable dependency of the shape of the gravitational field on matter can result in a deformation of the gravitational field. In other words, we find mass curving space. The only alternative would be a gravitational field which was flat in all circumstances. However, such a structure would effectively perform a function identical to Newton's absolute space – and we have already refuted the notion of absolute space.

So, as we discussed in Chapter Two, once again we appear to have a side effect caused by Nature being fundamentally unable to access absolutes. Nature does the best it can with the tools it has available, and, for the most part, we do not notice any problems – space appears flat to us. Light appears to travel in straight lines. We only notice the side effects caused by the lack of absolutes when Nature is pushed to extremes, in this case, in the presence of extreme masses. The inevitable side effect in this case is a curving of space around those large masses. For example, if we look closely we will find that the direction of light is bent slightly as it travels round stars due to the curvature of space. We interpret this curvature of space – which appears to attract objects to large masses – as the force of gravity. Hence, gravity itself is shown to be a side effect of the lack of absolutes in the universe.

It might appear strange to think of gravity as a counter-intuitive "glitch", much like quantum mechanical behaviour. However, this is only because we have spent our entire lives glued to the surface of a mass composed of 6×10^{24} kg! It seems to be very normal to us to be sucked at great force towards large masses. But although curvature of space appears very normal to us, most space in the universe is approximately "flat" (uncurved). Though we inhabit a very atypical area of the universe, we should not let this bias our way of thinking. The existence of gravity can be thought of, most certainly, as an unusual "glitch".

4

TIME IS NOT A CLOCK

Let us go back to 1905. We find a young Albert Einstein working as an assistant examiner in Bern patent office. Einstein had suffered recent disappointment. He had just finished a four year mathematics and teaching programme at the Zurich Polytechnic, but had failed to obtain an academic position. However, Einstein was not disheartened, and continued his personal research, despite the setbacks. His time at the patent office is often pictured as a dreary waste of time for the greatest mind in modern science, though it is more likely his job provided him with the financial support and freedom to pursue his own agenda which he likely would not have enjoyed in university. Einstein's friend, Abraham Pais, described Einstein's time at the patent office as "the closest he would ever come to paradise on earth".

Einstein spent his spare time trying to solve the most pressing physics problems of the day. In particular, he tried to solve a conundrum which had been troubling him from childhood: what would you see if you moved fast enough to catch up with a ray of light?

The Scottish physicist James Clerk Maxwell had just shown that light should be considered as being a type of electromagnetic wave. This posed a dilemma for physics. All types of known waves took the form of a disturbance in some type of underlying medium. For example, waves on the sea are a form of disturbance on water, while sound waves are a form of disturbance in air. But what medium carried light? Light was a disturbance of which medium? Physicists at the time called the proposed medium the *ether*. It was proposed that light took the form of a travelling wave of disturbance through the ether.

What form would this ether take? Sound waves require air, and there is obviously no air in space, but light can reach us from the distant stars. So the ether had to be a mysterious, invisible background substance which pervades the entire universe.

But Einstein was unhappy with the proposed ether idea. Galilean invariance stated that the laws of motion were the same in any uniformly moving environment – you could not conduct an experiment to determine if you were moving or not. However, if we introduce the idea of the ether it would be possible to perform an experiment to measure the speed of light and hence determine if you were stationary or moving: only if you were stationary with respect to the ether would you get the correct measure of the speed of light. So in many ways, the ether would have to play the role of Newton's absolute space – a form of absolute coordinate system which spread throughout space. But we have already refuted the notion of absolute space.

It did not seem right to Einstein that the laws of motion were the same in any uniformly-moving environment, but the laws of electromagnetism should be different (different observers obtaining different values for the velocity of light). Furthermore, in another blow to the ether idea, Albert Michelson and Edward Morley performed a famous experiment in 1887 which failed to detect any trace of the ether.

In a brilliant flash of inspiration, Einstein decided that the only way out of this dilemma was to reject the notion of the ether and extend the principle of Galilean invariance to the whole of physics. He proposed his principle of relativity: all the laws of the physics (not just motion) are the same in all uniformly moving environments. It might seem quite an innocuous principle, but it had tremendous implications. That simple principle formed the basis of the theory of special relativity, and from that point on, our concept of space and time would be forever altered.

Once again, we find we can derive this result from our fundamental principle. Considering our discussion of the two floating astronauts in the previous chapter, we saw that Nature could not distinguish which astronaut was stationary and which was moving. This is due to Nature being unable to access any absolute reference scale "outside the universe" (remember: there is nothing outside the universe). No astronaut can claim to be preferred. The experience of both astronauts is identical. We saw that this principle is called Galilean invariance. It is due to this inability of Nature to access **any** absolute reference axes outside the universe which means that Galilean invariance must be extended to **all** of physics. This leads to the principle of relativity: all the laws of physics are the same for all observers, no astronaut can claim to be preferred, the experience of both astronauts is identical. This means that both astronauts must measure the same value for the speed of light.

We shall now see how this invariance in the measured speed of light for all observers — regardless of their velocity — results in time dilation.

What time is now?

We are going to reveal the implications of Einstein's great insight by performing one of his famous thought experiments. Let us imagine we are riding on a train, and we have the latest technological marvel on our table: a light clock (there is no such thing in reality – it is purely a construct of our imagination). A light clock consists of two mirrors, and a pulse of light bounces backwards and forwards between the two mirrors at regular intervals. So this is what the light clock would look like if you were on the train:

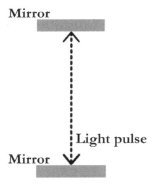

It is, in principle, easy to use the light clock to measure time as we know the distance between the two mirrors and we know the speed of light. What is more, the light clock is wonderfully accurate as we know the speed of light is a constant.

However, we now ask the question, what would the light clock look like to someone standing on the platform at a station as the train passes by? As the train is moving, anyone standing on the platform would see the pulse of light having to travel further between each pulse as the pulse now has to travel diagonally. So, for an observer standing on the platform, this is what the light clock would look like at three separate moments:

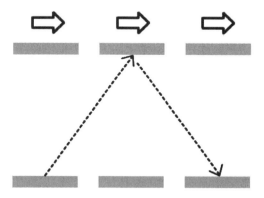

So, for each "tick" of the light clock, it appears to the observer on the platform that the pulse of light has to travel a greater distance than is the case for the passenger on the train. But the principle of relativity tells us that the speed of light is measured to be the same speed in every uniformly moving environment. That would appear to indicate that the observer standing on the platform would see a longer time interval between ticks than the passenger on the train would see. In other words, the observer on the platform sees the clock on the train running slow.

As bizarre as it might seem, this is precisely what happens. But it is vital to stress at this point that this result has nothing to do with light or light clocks in particular. What applies to the light clock applies to any clock, or indeed any physical process. It is not only the light clock which runs slow, but – according to the observer on the platform – it is time itself which runs slow on the train!

This phenomenon of the slowing-down of time for observers in relative motion is called *time dilation*.

However, the principle of time dilation is often badly expressed as "moving clocks run slow". Such a statement is relativistically flawed, and only adds to confusion. Who is to say which clock is stationary, and which is moving? Remember, there is no such thing as absolute space, no fixed coordinate axes to determine which clock is stationary. It is true that an observer will see the clock of another observer – who is in relative motion – running slow. But exactly the same applies from the point of view of the other observer. Observer 1 will see the clock of observer 2 running slow, and, because of the symmetry of the situation, observer 2 will see the clock of observer 1 running slow as well.

"How can this be?", you might ask. "How can both observers see the clock of the other observer running slow?" Without going into too much detail, the answer is that we have to bring the clocks together for direct comparison. If both observers decelerate to a stop in the same way, and then come together again (in a symmetrical manner), then when both clocks are stationary in the same reference frame and are compared, they will both say the same time.

So, instead of summarizing time dilation as "moving clocks run slow", be careful to say "a clock which is moving relative to an observer will run slow relative to the observer's clock."

This is no mere theorizing. In 1971, Joseph Hafale and Richard Keating flew four highly-accurate caesium-beam atomic clocks twice around the world on a jet liner, first eastward, then westward. When they got back to base, they compared the time on the clock with the time on a clock which had remained at base. They found the clock which went around the world had run very slightly slower: just a few hundred billionths of a second less elapsed time, but this was perfectly in accordance with Einstein's theory.

The rate at which time passes appears to vary depending on the motion of the environment. It would appear that each observer, moving through the universe, has his own personal clock which tells his own personal time. How can this be?

Time is not a clock

Isaac Newton believed in the concept of absolute time, along very much the same lines as absolute space. Just as absolute space proposed one series of absolute coordinate axes for space, absolute time proposed one single measure of time for the entire universe. According to Newton, it is as if there exists an invisible clock, which only Nature can access. This clock forms the universal standard, and "ticks" at the same unvarying rate for all objects. According to Newton, all physical processes are controlled by this universal clock:

> *Absolute, true, and mathematical time, of itself and from its own nature, flows equably without relation to anything external.*[8]

However, once again, Gottfried Leibniz took issue with Newton's concept of absolute time in just the same way as he had attacked his concept of absolute space. As with the case of absolute space, Newton believed that if all the matter in the entire physical universe was removed, the universal clock would still exist, ticking-away the absolute time. However, similar to the argument regarding absolute space, we can see that if there is nothing outside the universe then

---

[8] Isaac Newton, *Principia*, 1687

there can be no universal clock outside the universe providing this absolute time measure.

As Einstein's discovery of time dilation revealed, in the absence of the universal standard of absolute time, it is as though every object in the universe is now free to use its own personal clock, ticking at a different rate from every other person's clock.

In the last chapter we saw how the lack of absolutes in space meant that Nature was unable to distinguish between objects which had been spatially translated. This led to a symmetry: if you performed the same experiment but moved it six feet to the left, you would get the same result. But if we see symmetry in space due to lack of absolutes, then we would surely expect to see a similar symmetry in time. And this is precisely what we do see. Just as we found spatial invariance, we now find time translation invariance: if you shift your entire experiment in time (perform the same experiment at a later date) then you will get precisely the same result.

Once again we find a symmetry caused by Nature's fundamental inability to distinguish between two physical situations.

Spacetime

Let us now consider another implication of the principle of relativity. Imagine in the precise centre of a carriage on a moving train there is a lamp standing on a table. When the lamp is turned on, an observer riding in the carriage will obviously see the light beam hit the front and rear of the carriage at precisely the same time.

So the following diagram shows the situation as seen by an observer riding in the carriage. As shown in the diagram, the observer sees the emitted light strike the front and the rear of the carriage at the same time:

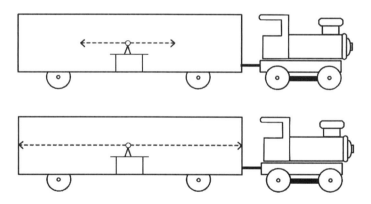

However, the situation is different for an observer standing on a platform watching the train pass by. According to the principle of relativity, all observers must get the same reading for the speed of light. So the observer on the platform will measure the same speed of light from the lamp on the train both in the forward and backward direction – despite the forward movement of the train.

In the time it takes the light to reach the ends of the carriage, the observer on the platform will see the rear of the carriage advance in the forward direction, due to the forward motion of the train. This has the effect that the light will reach the rear of the carriage first.

So the following diagram shows the situation for an observer standing on the platform. As shown in the diagram, the observer sees the emitted light strike the rear of the carriage first:

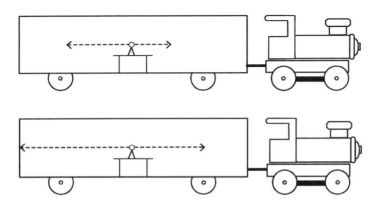

This result shows that events which appear simultaneous for one observer are not necessarily simultaneous for another observer who is in relative motion. This extraordinary, counter-intuitive behaviour only becomes noticeable at speeds close to the speed of light. As was explained in Chapter Two, at human scales and at human speeds we do not experience the strange glitches of quantum mechanics and relativity. It is only when Nature is pushed to the limit that the complete lack of absolutes available to Nature starts to show, and we experience the strange, counter-intuitive side effects.

This so-called *relativity of simultaneity* means that the ordering of events can be different depending on your viewpoint. I might see event A happening before event B, whereas you might see event B happening before event A, depending on our relative motion. As a result of this, when we are drawing events in space, it therefore makes no sense to attempt to draw all the events at a particular specified time, as different observers will disagree as to which events occurred at that time. The only way to unambiguously draw events in space is to add an extra time dimension onto a diagram of space, and then draw events in a block which incorporates both space and time. The resultant block of space and time is called *spacetime*.

The following diagram shows events (the black dots) in a block of spacetime (only two dimensions of space, *x* and *y*, are shown):

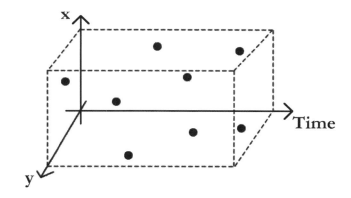

Spacetime is believed to be the most accurate representation of space and time. When considering the structure of space and time, we should treat the time dimension as being combined with the three spatial dimensions to form a four-dimensional spacetime. Although it sounds rather far-fetched, we in fact treat time as a fourth dimension every day. For example, when we arrange to meet someone, we not only have to say where (in space) we want to meet them but we also have to say when (in time) we want to meet them. So the meeting event is specified by four pieces of information: the three dimensions of space, and the one dimension of time.

It is hard for us to imagine time as the fourth dimension because, whereas objects are free to move in any direction in the space dimensions, they only move forwards in the time dimension. This makes it especially difficult for us to imagine time as a dimension as – psychologically – it seems so very different from the space dimensions. And, while we can visualize three orthogonal axes, it is simply impossible to visualize four orthogonal axes. However, if we overcome our psychological objections we find that time does, indeed, behave very much like the dimensions of space.

As we move about in space, we are also moving forward in time. This means we effectively navigate a path through spacetime called a *world line*. This is an important concept and we will be returning to this idea later:

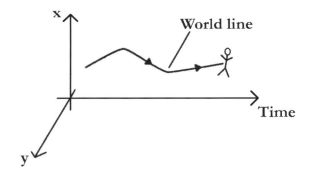

It was Hermann Minkowski, one of Einstein's teachers, who first conceived of four-dimensional spacetime. Minkowski famously said: "Henceforth space by itself, and time by itself, are doomed to fade away into mere shadows, and only a union of the two will preserve an independent reality."

Invariants

In our light clock thought experiment, we have already seen how Einstein showed that time would pass slower for an observer who is in relative motion. It would appear that space and time are very much relative quantities, and the measure you get for these quantities depends on your viewpoint. However, when we combine space with time to form spacetime we find we get some joint quantities which do not depend on your viewpoint. Such quantities are called *invariants.*

The first invariant we will consider is the distance travelled by an observer in spacetime. In the following equation, the spacetime distance, s, is equated with the length of time measured by an observer, t, the spatial distance travelled by the observer, x, and the speed of light, c:

$$s^2 = (ct)^2 - x^2$$

It so happens that the spacetime distance travelled by an observer – as calculated by this equation – is invariant, i.e., all observers will agree on this distance.

What are the implications of this invariant spacetime distance? In order to answer that, let us consider two observers. One observer stays on planet Earth, while the other observer decides he wants to be an astronaut and announces his intent to travel to the edge of the solar system before returning to Earth. Both observers synchronize their watches before the astronaut starts on his journey. When the astronaut returns to Earth, the two observers compare their watches again.

You will note that the value for x^2 in the spacetime distance equation is negative. This means that the observer which covered the greater spatial distance will have the smaller distance in spacetime. In other words, the spacetime distance – which represents the amount of time experienced by an observer – will be smaller for the astronaut which went on the round-trip to the edge of the solar system. Less time will have passed for the astronaut than for the observer who remained on Earth.

This is precisely in accordance with Einstein's theory of time dilation: the moving astronaut's clock runs slow compared to the clock of the observer who remained on Earth. If the astronaut travelled far enough and fast enough he would have returned to find his friends had aged considerably, while he remained young.

What else can the spacetime distance equation tell us? Well, if our spatial distance, x, is zero, i.e., if we do not move through space, then the spacetime distance we travel is equal to ct. In other words, just by standing still we are all travelling through spacetime at the speed of c, the speed of light!

How absolutely bizarre, you might think. You might think that when you are standing still or sitting down, you do not feel as if you are travelling at all. Well, you do have the sensation, really: you feel as if you are travelling in time. You feel a kind of motion of the present moment, as the future becomes the past. Just by sitting down, you are travelling through spacetime at the speed of light (do not take this as an excuse to do less exercise).

So how should we interpret forward motion in the time dimension? It means you are getting older! That is how to interpret motion in the time dimension: the object doing the travelling in the time dimension gets older

This invariant quantity – that everyone is moving in spacetime at the same speed – gives us a different way to look at Einstein's theory of special relativity. When we combine space and time into spacetime, we find that all observers move through spacetime at the same speed. This means that if you see another object is moving at high speed relative to you, its speed through space is greater than yours. Hence, its speed through time must be less than yours. This is what you will find when you compare your clock with a moving clock: as the moving clock is travelling faster through space, it is travelling slower through time, i.e., time runs slower for the moving clock relative to your clock. Actually, if an object could move through space at the speed of light (as massless photons can), they would not move through time at all – all their speed would be motion through space, with no speed left to travel through time. For this reason, photons do not experience the passage of time at all.

As Brian Greene says in his book *The Elegant Universe*: "We now see that time slows down when an object moves relative to us because this diverts some of its motion through time into motion through space." This is time dilation expressed from the point of view of invariants: everyone's speed through spacetime is the same invariant quantity. In fact, Einstein did not originally want his theory to be called "relativity", he wanted it to be called "invariance", precisely for this reason.

So we now see that the speed of light is not actually the cosmic "speed limit" in the space dimensions it is usually portrayed to be. It is, in fact, the speed at which **everything** is travelling – in four-dimensional spacetime! So the next time that someone in the pub tells you that no one can travel at the speed of light, you can tell him that he is travelling at the speed of light right now!

The speed of light is no speed limit – it is the universal speed. Everything travels at the speed of light.

Invariants – not absolutes

At the end of the last chapter, we discovered that acceleration is defined to be an "absolute", which explained how Nature could distinguish between uniform motion and motion undergoing acceleration. In this section, we are going to explore precisely what is meant by the term "invariant" and compare it to the similar term in common usage "absolute". This has important repercussions for the main theory presented in this book.

In order to do this, let us imagine a group of objects which are moving about in space at different speeds, and each one is travelling in a straight line (i.e., this represents uniform motion). Also, let us say that each of these objects has a limited lifespan, for example, ten days. This means after each object appears in the universe, it exists for ten days according to its own clock, and then completely vanishes from the universe.

From our previous discussion, we know that all objects travel at the speed of light in spacetime, so if we draw the world lines of these moving objects we will find them to be orientated differently (as some objects move through space at different speeds relative to other objects), but the world lines will all be the same length:

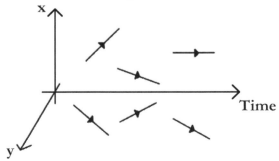

The world lines are the same length because their speeds in spacetime are invariant, and the lifetimes of the objects are the same. Hence, we have a quantity which is invariant: the length of the world lines. But does this mean that this length should be considered an "absolute"?

To answer this, remember back to Chapter Two when we considered the timber factory. If you remember, Nature was in charge of running the timber factory, and had the task of producing 1,000 metre-long planks of wood. As Nature was unable to access any absolute measures outside the closed system of the timber factory (it was unable to access the platinum bar in Paris), it had to pick one plank at random, and cut all the other planks to the same length relative to that randomly-chosen plank.

We now find the situation with the world lines to be a direct analogy. The lines are all the same length. Nature can ensure these lines are the same length, and it can do this without having to access any absolute measures outside the universe. The lengths of these world lines might be described as an "absolute", but we can see that they are really **relative** lengths: the lines are all the same length relative to each other.

The term "absolute" is used because all observers can agree that the lengths are the same value – the value does not depend on your frame of reference. But this use of the term "absolute" is unfortunate as we can see that it really refers to a relative quantity. It is much better to use the term "invariant" to describe the lengths of the world lines. Einstein was quite correct: "invariance" was a much better name for his theory.

Let us now move on to consider the case of accelerated motion. Remember back to the last chapter and our discussion of Mach's principle. We were wondering how an astronaut floating in space in a universe which had all other matter removed would ever know he was spinning. In such a situation, how could Nature ever distinguish between uniform motion and accelerated motion? The answer we used at the time was that "acceleration is an absolute". But, if this is the case, then we will have to consider this claim a lot more closely.

What we find is that Nature can view the whole of time and space – no information **within** the universe is hidden from Nature. This means Nature can see all time, and, therefore, Nature can see whole world lines of objects as they travel through time. Nature can see that the world line of an object in uniform motion is a straight line, whereas the world line of an object in accelerated motion will have a variable velocity and its world line will therefore be a curved line:

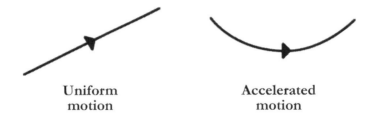

**Uniform   Accelerated
motion   motion**

As Brian Greene says in *The Fabric of the Cosmos*: "Geometrical shapes of trajectories in spacetime provide the absolute standard that determines whether something is accelerating."

As Nature can see the entire world line of the astronaut, it can detect if this world line is straight or curved. Just look at the two world lines in the diagram on the previous page – they are clearly distinguishable without any need to refer to coordinate axes: one is straight, while the other is curved. If someone handed you a straight piece of metal and a curved piece of metal, you could clearly distinguish the two pieces. And Nature can do exactly the same thing with world lines: it can distinguish uniform motion from accelerated motion just by considering the shape of the world lines.

Bear in mind, though, that it is the geometry of space which plays the vital factor in determining what is straight and what is curved. And, as we discussed in the previous chapter, it is the gravitational field which plays a role which is synonymous with space. So Nature is really detecting the curvature of the object's world line **relative** to the gravitational field. So really, rather than considering acceleration to be an absolute, what we find we have here is another relative measure: acceleration is curvature relative to the gravitational field.

However, the gravitational field itself is also a relational structure. It has no dependence on any absolute coordinate axes – it just depends on the masses in the universe. In the previous chapter we discussed how the presence of other masses had the effect of warping the gravitational field. The force of gravity an object feels is therefore relative to the other objects in the universe. In this sense, the gravitational field acts as a kind of "intermediary" between masses: masses which are moving through space, and masses which are warping space.

Returning to consider our isolated astronaut who is the only object in the universe in which all other matter has been removed, how can this new picture explain how Nature knows the astronaut is spinning, i.e., accelerating? With no other masses in the universe, Nature can ascribe no particular shape to the gravitational field. The gravitational field would then be uniform over the entire universe (it would be said to be "flat"). Hence, the only factor determining the acceleration of the astronaut is the curvature of his world line in a flat spacetime. As we have just discussed, Nature can clearly distinguish the difference between a straight world line and a curved world line. Effectively, Nature is determining the curvature of the world line relative to the rest of the world line. So the curvature of the line is being measured **relative to itself**. So even in a universe in which all the other matter had been removed, Nature could still tell if an astronaut is spinning.

When we have considered acceleration in this discussion, all we have found is a series of relative measures. Acceleration is curvature relative to the gravitational field, the gravitational field is itself a fully relational structure shaped by the other masses in the universe, and, in empty space with no other masses, acceleration is curvature of the world line relative to itself. So everything about acceleration is actually relative. In that case, why is it so often described as being an "absolute"?

It is commonly stated that acceleration is an absolute because all observers in all frames of reference will agree on the value of the acceleration. But, does this make for an absolute reference in the truest sense of the word "absolute"? As we saw in the case of the timber factory, all the planks of wood might have been the same size, but they could be **any** size in the absolute sense (i.e., when the customer constructs his house and finds it to be a random size). Mutual agreement about a value by all observers in a closed system is not sufficient to make a true absolute. In order to be a true absolute, there has to be a standard of reference outside the closed system. But there can be no standard of reference outside the closed system of the universe.

In all our discussions, we have found that all measures which might be considered absolutes are actually relative measures, measures taken relative to other objects within the universe. This is an inevitable consequence of there being nothing outside the universe. Just by considering the obvious logic of that situation, we can see that Nature fundamentally unable to access any standards of reference outside the universe. Hence, there can be no true absolutes within the universe.

The proposal of this book is that there are absolutely no absolutes, in the truest sense of the word "absolute". We can sum this up by saying that **reality is relative**.

We will see throughout the remainder of this book that this principle – that reality is relative – is not generally realised by many physicists (or, at least, it is not reflected in their work). This leads to confusion and apparent paradoxes – especially, as we shall see, in matters concerning time and quantum mechanics. As soon as we switch to considering the universe as being a purely relative entity, this confusion disappears, and the paradoxes are revealed to be purely a product of using an inaccurate model of the universe.

5

THE BLOCK UNIVERSE

In this chapter we will continue to consider the implications of there being no absolute axes of time outside the universe. The implications of this turn out to be so remarkable that this discussion may very well change your conception of time forever – I know it had that effect on me. I genuinely view time, and events in my life, in a different light now that I understand the full implications of the block universe model.

The path we will take will be purely logical, and our conclusions will be clear and unarguable. This makes it all the more remarkable that many professional physicists seem unaware (or choose to ignore) the implications of basic logic when applied to time. As Lee Smolin says in his book *Three Roads to Quantum Gravity*: "There are unfortunately not a few good professional physicists who still think about the world as if space and time had an absolute meaning."

It is time to spread a bit of logical light ...

How fast does time flow?

In the last chapter, we discovered how time is a relative notion. This is an inevitable consequence of our principle that there is "nothing outside the universe" – different observers in the universe can effectively have their own "clocks". Time appears to pass at different rates for different observers, relative to other observers.

Then, in our discussion of the relativity of simultaneity, we found it made no sense to attempt to draw all the events at a particular specified time, as different observers will disagree as to which events occurred at that time. We found that the only way to unambiguously draw events in space is to add an extra time dimension onto the diagram of space, and then draw events in a block which incorporates both space and time. Thus, we represent spacetime graphically as a block, and this is the most accurate way to think about space and time.

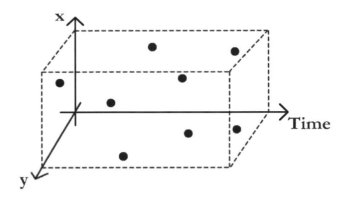

But there is another, quite fascinating, logical way to consider time which also suggests that we should consider spacetime as a block. We have all seen objects changing with time. We see cars driving down a road, or birds flying in the sky. In slightly more formal terms, what does this motion actually represent? It represents a change in the position of an object with respect to time. As time progresses, the object's position alters. We interpret this change in position as motion. Some objects move slowly with respect to time, some objects move quickly with respect to time.

Bearing this in mind, in 1951 the American philosopher D.C. Williams asked a simple question: "How fast does time itself move?"[9] This might seem like a fairly innocuous statement, but it is far from easy to answer. We have discussed how we see objects moving with respect to time, but if time itself is moving, then is it moving with respect to itself? This sounds nonsensical. Maybe we shall have to elaborate on what the movement of time actually means.

We all have an intuitive feeling of the passage of time. We feel the immediacy of a "now" moment as if it is the only real moment, we remember the past as "somewhere we have been", and we look to the future as "somewhere we are going". In this sense, it feels as if our "now" moment is moving through time, from the past to the future. The movement of the "now" point has the effect of turning the uncertain and malleable future into the fixed and unchangeable past.

So, if we are going to answer the question of "How fast does time itself move?" then it would appear we need to consider how fast the "now" point moves. If the "now" moves then it must move with respect to some time reference. So is time moving with respect to itself? Surely not. To say "Time moves at the rate of one second per second" is meaningless. Rather, the rate of time flow would

---

[9] Donald C. Williams, *The Myth of Passage*, Journal of Philosophy 48 (15):457-472

have to be measured with respect to some secondary, external time reference. However, as we have discussed in earlier chapters, we know that there is no clock outside the universe, so there could not be any such external time reference. It is simply logically impossible for there to be a moving "now". The only logical conclusion we can draw is that time does not flow!

So what is the alternative? The alternative is to consider a universe in which all of time is laid-out (just as the space dimension is laid-out), and there is no moving "now". We would imagine this as a solid block of unchanging spacetime, containing all the events of all time. As there is no special "now", there is no distinction between past and future. Amazingly, this means that **all times are equally real**. The future and the past are just as real as the present!

If all times are equally real, then the times and events you remember earlier in your life are just as real as now: the current moment can have no particular claim to be a special moment in time. So, you remember that moment as a child when you fell off your bike and grazed your knee? That moment is just as real as the current moment – it can be said to exist just as much as the current moment. All times exist, all times are real.

This principle forms the *tenseless* theory of time. It is also called the *block universe* model because all of spacetime can be viewed as being laid-out as an unchanging, four-dimensional block. The block universe model can be considered the orthodox model of spacetime as it has its roots in special relativity and basic logic, and is the model of spacetime adopted by the vast majority of physicists.

I first became aware of the full, extraordinary implications of the block universe model in 2006 when I read a superb *Scientific American* article by Paul Davies entitled *That Mysterious Flow*. I suspect that for a lot of people – not just me – the article was a revelation. Paul Davies presented the argument against the conventional, intuitive notion of a moving "now" in a way which really made you stand up and take notice:

> It makes sense to talk about the movement of a physical object, such as an arrow through space, by gauging how its location varies with time. But what meaning can be attached to the movement of time itself? Relative to what does it move? Whereas other types of motion relate one physical process to another, the putative flow of time relates time to itself. Posing the simple question "How fast does time pass?" exposes the absurdity of the very idea. The trivial answer "One second per second" tells us nothing at all. [10]

I reproduced Paul Davies's article on my website and got a tremendous response. The general consensus from the comments told me that the average scientific lay-reader was already well-acquainted with the idea of all of space and time being laid-out in a spacetime block (i.e., they were already well-acquainted with the notion of a four-dimensional spacetime – thanks to Einstein). However, they were still under the impression that there was a "now" moment which moved through this spacetime structure, turning the future into the past. Brian Greene also noted this general misunderstanding in his book *The Fabric of the Cosmos*: "A less than widely appreciated implication of Einstein's work is that special relativity really treats all times equally." It is this misunderstanding which I hoped to clarify – with the aid of Paul Davies's article. All of space and time is laid-out, yes, but there is no moving "now". All times are equally real.

---

[10] *That Mysterious Flow* by Paul Davies:
http://www.whatisreality.co.uk/reality_mysterious_flow.asp

According to the block universe scenario, the movement of time is just an illusion. "But", you might protest, "I clearly feel the current moment is special. The past seems distant and hazy, and the future is unknown. So the current moment is definitely special for me." Well, yes, but do not forget that this is precisely how you have felt at every other moment in your life. In 1980, you felt that the current moment was special, and the 1970s were the blurry past. In 1990 you felt that the current moment was special, and the 1980s were the blurry past. In 2000 you felt the current moment was special, and the 1990s were the blurry past. The way you are feeling now is precisely how you have felt at every other moment in your life. So, again, there is nothing to identify the current moment as special in any way.

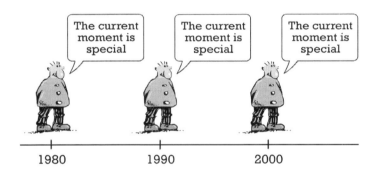

It is as if there are multiple copies of you in existence at every point in time, and they are all equally real. Remember the discussion about your world line in the previous chapter. That is the accurate representation of reality: you exist as a world line across all times. Just considering yourself at the current moment does not reflect the true reality.

Are we *The Sims*?

David Deutsch provides an excellent description of the block universe model in Chapter Eleven of his book *The Fabric of Reality*. In fact, he comprehensively demolishes the generally-held view of time in favour of the block universe view:

> *The reason why the common-sense theory of time is inherently mysterious is that it is inherently nonsensical. It is not just that it is factually inaccurate. We shall see that, even in its own terms, it does not make sense ... Common sense frequently turns out to be false, even badly false. But it is unusual for common sense to be nonsense in a matter of everyday experience. Yet that is what has happened here.*

Deutsch even considers the possibility that there might be an axis of time external to the universe by which we might answer the question "How fast does time flow?" Deutsch achieves this by considering the possibility that our universe might be a computer simulation, the simulation being performed by an advanced civilisation. If you have seen the movie *The Matrix*, you will be aware of a theory similar to this. In *The Matrix*, the advanced civilisation kept Keanu Reeves's character in a vat and fed his brain false stimuli to make it appear as if he was living in a completely false, simulated universe. The latest theories on this subject have attained a degree of scientific credibility, but they do not propose keeping anyone in a vat. Instead, these theories consider the possibility that our entire universe might be a simulated construct in a vast supercomputer run by an advanced civilisation (as if we were simulated characters in the computer game *The Sims*). The motivation behind such a simulation being just the same as why we enjoy playing games such as *The Sims*: for entertainment.

If we were in such a position, then there would be two axes of time: there would be our time axis as residents inside the simulated universe, and there would be another, entirely separate axis of time for the advanced civilisation performing the simulation – effectively outside our universe. The two axes of time need not run at the same rate, e.g., the advanced civilisation could be watching us go about our daily activities in slow motion.

It would then appear that we could provide an answer to the question "How fast does time flow?" We could say something like "Our time inside the universe flows at five seconds for each elapsed second of time outside the universe." In other words, although it is a nonsense to say that time flows at a rate of "one second per second", it would appear possible to say that time flows at a rate of "one second per external second, according to the time scale outside the universe."

Is this a way out of the logical certainty that we inhabit a block universe?

Well, no. Deutsch shoots down this theory as soon as he proposes it. He points to the fact that we would be just pushing the problem back from our simulated universe to the universe of the advanced civilisation. We may have avoided the problem "How fast does time flow?" in our simulated universe, but we are now left with the question "How fast does time flow?" in the universe of the advanced civilisation. We are no better off.

There is no way out: it is a logical certainty that we inhabit a block universe.

Eternal life

It might come as a surprise that this orthodox block universe view of time leads us to conclude that we possess a form of eternal life! This is a consequence of the principle that in the block universe model all periods of time are equally real. If a loved one dies, you might take some comfort from the knowledge that this period of time in which your loved one is dead has, in fact, no greater reality than the time when your loved one was alive. According to physics, it is just as valid to consider your loved one as alive as it is to consider them dead!

Einstein took comfort from this knowledge when his lifelong friend Michele Besso died. He wrote a letter consoling Besso's family: "Now he has departed from this strange world a little ahead of me. That means nothing. People like us, who believe in physics, know that the distinction between past, present, and future is only a stubbornly persistent illusion."

Of course, the flip-side is that you're already dead!

Free will

The idea that all of space and time is laid-out in one unchanging block might appear unsavoury to some people as it appears to deny the possibility of free will. This is because it appears we are unable to change the future, which is written in stone. This is not the first time that science has appeared to deny the possibility of free will: Newton's laws of motion seemed to imply a deterministic "clockwork universe", in which, once the initial conditions are set, the universe proceeds along a fixed path which cannot be altered. The arrival of quantum mechanics, though, implied that the universe is not totally deterministic. Quantum mechanics only supplies the probability of an event occurring. It is as if quantum mechanics decides which way the world is going to work by throwing a dice.

But does quantum mechanics actually provide you with more free will than Newton's clockwork universe? I fail to see how having your decisions controlled by a random dice throw somehow gives you more control than having your decisions controlled by a deterministic mechanism. It seems that the proponents of free will try to deny that their thought processes are controlled by fundamental physical processes at all, as somehow their thoughts are "above all that". Descartes, for example, believed that the human mind did not have to follow the physical laws of the universe. This is surely not the case. However you define free will it is surely controlled by the laws of Nature and fundamental physics.

The main problem is that "free will" has no proper scientific definition, and until it is given a proper definition it is simply impossible to analyse it scientifically to see if it is present or not. In the rather un-scientific way that it is normally described, free will is "the ability to make decisions", and that is something we all clearly have. Whether the underlying universe is deterministic, or random, or a block universe, it makes no difference: we all have the freedom to make decisions. This is because "the ability to make decisions" is, again, not sufficiently defined scientifically. In this un-scientific sense, the only way we could ever lose this ability to make decisions is if a drug was put in our drink to turn us into a zombie-like, suggestible state. So, in the absence of someone drugging your drink, I can definitely say, yes, you obviously have free will.

The arrow of time

The block universe model tells us that all times are equally real, and the passage of time is just an illusion. It certainly feels like a very strong illusion, which is why it is sometimes so hard to accept the reality of the block universe. So what is causing this illusion of the passage of time?

The intuitive feeling we all experience is of a moving "now" which turns the future into the past. This seems to indicate a clear directionality of the movement of the "now" from the past into the future. There is a clear asymmetry in the time direction, most notably we can remember the past but the future is completely unknown. This asymmetry in our memory seems to be the main reason why we feel we are in motion in the forward time direction. Because we can remember the past, we feel that it is somewhere we have already been, and because we have no knowledge of the future we feel it is somewhere we have yet to visit. Hence, despite what the block universe model tells us, we feel we are moving in time – we feel the passage of time.

So the asymmetry of the time axis causes an apparent directionality to time. But what is the cause of this asymmetry? What is the cause of the so-called *arrow of time*?

The asymmetry is not just in our minds and memory. As we look around our world, we see wine glasses breaking, but we never see broken pieces of glass forming a wine glass. We see eggs breaking, but we never see a broken egg reforming. So there is asymmetry in the physical world along the time axis. This is particularly puzzling when you consider that physicists believe that almost all physical processes are reversible. There should be no fundamental reason why objects behave differently in the forward time direction to the backward time direction.

Instead of considering time, we should consider which processes appear to be irreversible. We should consider what a breaking wine glass and a breaking egg have in common. In both these cases, we see an increase in disorder. It would appear that it is much more likely that a system could change from a state of order into a state of disorder rather than vice versa. Indeed, this does appear to be the reason why we see an arrow of time.

The scientific term for the amount of disorder in a system is *entropy*. It is known that the entropy of a closed system increases with time, i.e., a system will gradually become more disordered over time. Hence, we see cars rusting (i.e., their molecules become more disordered), but we do not see rusting cars becoming new again. This principle that the entropy of a closed system increases with time is called the *second law of thermodynamics*.

The reason for this increase in entropy can be seen from a purely probabilistic argument: a system will have many more possible disordered states than ordered states, so a system which changes state randomly will most likely move to a more disordered state. It is really just a matter of likelihood. For this reason, the second "law" of thermodynamics is not really a "law" at all, certainly not an unbreakable law on the same basis as other physical laws – it is a statistical principle. In fact, it might be possible for a room full of randomly-distributed particles to re-order itself quite by chance so that all the particles end up in one corner of the room – it would just be incredibly unlikely!

While the second "law" of thermodynamics is "just" a statistical principle, it is a mightily powerful statistical principle! This is because the basis of the second law – that "disorder will increase" – seems so obvious, and seems to appeal to a fundamental, platonic principle of mathematics. For this reason, the second law manages to appear even more fundamental and unbreakable than the other physical laws, some of which (for example, the amount of electric charge on an electron) seem rather arbitrary in comparison. This fundamental strength of the second law is described well by the astrophysicist Sir Arthur Eddington:

> *If someone points out to you that your pet theory of the universe is in disagreement with Maxwell's equations, then so much the worse for Maxwell's equations. If it is found to be contradicted by observation – well, these experimentalists do bungle things sometimes. But if your theory is found to be against the second law of thermodynamics I can offer you no hope; there is nothing for it but to collapse in deepest humiliation.*

I would consider this principle – that disorder will tend to increase – to be so strong that it would be "obviously true" in all conceivable universes. For this reason, I would consider the second law of thermodynamics to be one of those rarest of things: a true fundamental principle, a principle with the potential to be one of the foundation stones of the laws of Nature.

Ex nihilo solutions

To my mind, the disconnect between theoretical physics and basic logic about the subject of time is greatest in the field of cosmology. Cosmology is the study of the entire universe, including its origin. One of the most popular fields of study attempts to explain the origin of the universe by invoking quantum theory. The resultant theory is known as *quantum cosmology*.

According to quantum theory, there exists some unavoidable uncertainties at the base of reality. It turns out that the more accurately we measure a particular property of a particle, the less accurately we can know a different particular property. For example, the more accurately we measure the momentum of a particle, the less accurately we can know its position. One way of looking at this is to imagine that when we measure the properties of a particle we inevitably have to interfere with it in some way, for example, by forcing it to collide with another particle. This interference inevitably changes the property of the particle under investigation – by modifying its velocity, for example. This fundamentally limits the accuracy of the measurement.

Such pairs of properties – such as momentum and position – which behave in this way are called *conjugate variables*. The theory which expresses this uncertainty in conjugate variable values is called the *Heisenberg uncertainty principle*.

According to the uncertainty principle, another pair of conjugate variables is energy and time. This means there is a limitation on how accurately we can know the energy of a system at a given moment in time. Incredibly, this means that for a short period of time, there can be energy produced in a vacuum in which all the matter has been removed. It is as if we can get something out of nothing! Even more incredibly, according to the mass-energy equivalence described by $E = mc^2$, this vacuum energy can be enough to produce a particle! It is as though the energy is "borrowed" from the vacuum, albeit for a very short period of time. These so-called *virtual particles* can appear out of nothing.

Quantum cosmology considers the possibility that one of these quantum fluctuations could have expanded extremely rapidly in a process known as *inflation*. The suggestion is that one of these fluctuations could have become our universe, blowing up like a balloon just a tiny fraction of a second after the big bang. Our universe would then have emerged from nothing! Theories which propose that our universe may have emerged from nothing in this manner are called *ex nihilo* solutions (*ex nihilo* being Latin for "out of nothing").

An *ex nihilo* solution which is very popular at the moment is based on the idea that the total energy of the universe can be calculated to be zero. That would be the only kind of universe that could come from nothing. The idea of the total energy of the universe being nothing might seem a strange idea as the universe contains matter, and we know through $E = mc^2$ that matter can be converted to energy. So how can the sum total of the energy in the universe be zero? This is possible if gravitational energy is considered to be negative. This requires explanation.

If objects are separated to infinity they feel no gravitational pull between themselves, so the gravitational energy of the system is zero in that case. But when those objects were initially clumped together, you had to put energy into the system to pull them apart. So if you have to put energy into a system just to get to a zero energy situation, this means the energy of the system when those objects were clumped together must have been negative.

So gravitational energy can be considered negative, and, if you actually do the arithmetic for the universe, you can show that the negative gravitational energy exactly cancels the positive energy due to the masses in the universe.[11] Hence, the total energy of the universe is zero.

Once again, this result could actually be derived from our fundamental principle. If there is "nothing outside the universe" then it could never be possible to "stand outside" the universe and measure any value for the universe as a whole. This principle is expressed by Misner, Thorne, and Wheeler in their classic textbook *Gravitation*: "There is no

---

[11] See Curtis Menning's website *Calculation of the Energy of the Universe*: http://www.curtismenning.com/ZeroEnergyCalc.htm

such thing as the energy (or angular momentum, or charge) of a closed universe, according to general relativity, and this for a simple reason. To weigh something one needs a platform on which to stand to do the weighing." There can be no such platform outside the universe. Once again, we find that our fundamental principle ("there is nothing outside the universe") can be used to derive a remarkable proportion of the laws of Nature.

This result – that the total energy of the universe is zero – makes *ex nihilo* solutions very attractive. It really makes it appear as though the universe could just have emerged from nothing, and that is the reason why the universe exists at all. As Alan Guth, the discoverer of inflation, said: "It is rather fantastic to realise that the laws of physics can describe how everything was created in a random quantum fluctuation out of nothing."

But it is not all good news. Some questions have been raised as to whether it is valid to apply quantum reasoning – which is only ever used to describe behaviour inside the universe – to the universe itself. Our later analysis of quantum mechanics will appear to show that quantum mechanical behaviour arises from the interactions of objects within the universe. In which case, it would indeed be invalid to apply such reasoning to the universe itself.

But a more fundamental problem is that ... ***ex nihilo*** **solutions make no logical sense whatsoever!**

This conclusion is entirely a consequence of the block universe. In order to see this, first it has to be stressed that accepting the reality of the block universe is not an option. To disregard the implications of the block universe is not only to ignore the conclusions of special relativity, it is to ignore basic logic. Remember the logical argument based on "How fast does time pass?" It is logically absurd to state how fast time itself passes, and it is therefore logically absurd to consider we do not live in a block universe. We have a definite logical conclusion: the universe has a block universe structure.

And the implication of the block universe is that all times are equally real – no particular moment is preferred. The current moment of the universe is just as real as the first moments after the big bang. The entire spacetime block is laid-out as one unchanging structure. No time is dependent on any other time – all times are real, all times are equally real. All times exist.

So to suggest that in some way that a particular moment of time, such as the time of the big bang, is somehow more important, more real, than other times is a nonsense. The last moments of the universe are just as real as the moment of the big bang. And to suggest that the events of any particular moment are responsible for the existence of the universe is complete and utter logical nonsense! No time depends on the existence of any other time. The universe is one unchanging block of spacetime. The last moments of the universe are just as important for the existence of the universe as is the time of the big bang.

This is surely an incredible conclusion. It has been derived logically, and so it is not open to rational argument. As such, it should be regarded as orthodox physics. And yet, not only is this result ignored by a large proportion of physics theorists, it renders a large proportion of their research logically absurd.

It seems incredibly counter-intuitive to suggest the big bang is not the reason the universe exists, but this is what logic tells us. The big bang was a massively important event for objects **within** the universe. It explains why the universe takes the form it does. But it tells us nothing about why the universe exists at all.

Ex nihilo solutions make no logical sense. The events of the big bang are not the reason the universe exists. The universe exists as an unchanging spacetime block, the form of which is described by Stephen Hawking in this extract from his book *A Brief History of Time*: "If the universe is really completely self-contained, having no boundary or edge, it would have neither beginning nor end: it would simply **be**."

6

QUANTUM REALITY

We are now approximately halfway through this book. As we have concentrated mainly on relativity so far, this would seem like a good point to turn our attention to quantum mechanics.

As far as the general public is concerned, quantum mechanics has a bad rep. It is perceived as being the ultimate in inaccessible, esoteric theoretical physics, only to be understood by the likes of Sheldon from TV's *Big Bang Theory*.[12] I think this is a tremendous shame as the basic principles of quantum mechanics are really remarkably straightforward. Perhaps physicists must bear part of the blame for not communicating this inherent simplicity to the general public. I do take a fair bit of personal pride in going the other way, identifying the fundamental principles which underlie a subject, and then expressing those principles as simply as possible.

---

[12] A string theorist.

I have always wanted to present the principles of quantum mechanics to a general audience in a form which differs from the approach of most popular science books. The general approach of most books is to start by describing the most bizarre and counter-intuitive examples of quantum mechanical behaviour. I am not quite sure as to what is the intention of this approach – maybe to get you into thinking "Ooh, isn't that difficult and weird!" This approach appears rather self-defeating to my eyes.

I have always felt that a better approach is to emphasize the fundamental logic which underlies quantum mechanical behaviour, which really is quite straightforward and can be described in a few sentences. Quantum mechanical behaviour would only be bizarre if it did not apply its rules consistently. On the contrary, quantum mechanics is totally consistent in applying the same underlying logic to all areas of application, and so should not be considered bizarre – perhaps counter-intuitive might be a better phrase. The presentation of quantum mechanics in this book will emphasize the logic, and de-emphasize the bizarre.

Another bugbear of mine about the presentation of quantum mechanics in the popular literature is that all too often it just becomes a history lesson, parading out the life-stories and achievements of the pioneers of the subject in chronological order. The result can resemble an episode of TV's *This is Your Life* rather than a popular science book. As truly great and heroic the achievements of those pioneers were, the important message to convey is the principles – not the personalities. I am going to try to get through this chapter without describing the story of a single pioneer of quantum mechanics. If you are looking for a book on the history of the subject, there are plenty on the market for you to choose from.

The logic of quantum mechanics

In general, we use our knowledge of quantum mechanics to predict the behaviour of systems. In the most general sense, this means predicting the result we will get when we measure some aspect of the system. So, in essence, quantum mechanics is all about measurement.

In order to explain the logic of quantum mechanical measurements, we need to consider the results of just three experiments. The first experiment we will consider tells us about the wave nature of light. It was first performed at the beginning of the 19th century by Thomas Young.

At the time, Newton's views on light were dominant. Newton believed light was "corpuscular" (made of particles), this conclusion being largely based on the observation that light moved in straight lines – just like any other matter in inertial motion. Thomas Young realised that this corpuscular model could not explain effects such as diffraction, which could only be explained if light behaved like a wave. So Young proposed an experiment which he believed revealed the true, wavelike nature of light.

In Young's experiment, a light source is in front of a board. Two narrow slits are cut into the board. Light can only pass through these two slits, and the light which passes through the two slits illuminates a screen behind the board.

The two light rays from the two slits meet at the screen. Due to the wavelike nature of light, the peaks and troughs of the light waves interfere with each other at this point. This means that, if we consider a single point on the screen, if the light wave from one slot has a peak in its waveform at that point, and if the light wave from the other slot also has a peak, then this represents constructive interference and there will be a bright spot on the screen. However, if there is a

111

trough projected from the second slot, then this represents destructive interference: the trough from one slit cancels-out the peak from the other slit. As a result, there will be a dark point on the screen.

Because of the variations in distance between the two slots and points on the screen, the resultant pattern shows both constructive and destructive interference. The result is a pattern of bright and dark lines being projected on the screen. The diagram below shows the experimental set-up, and the pattern of dark and bright lines being projected on the screen:

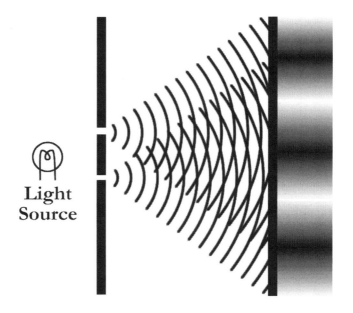

Light Source

So Young's experiment clearly showed that light behaves as a wave. However, the next experiment shows that this does not capture the full description of the behaviour of light. The second experimental result we shall consider is the photoelectric effect, and it is here that quantum effects shall start to become noticeable.

Next time you are in a queue at the supermarket, take note that you are in the presence of the experiment which launched quantum theory. When you put your groceries on the rubber conveyor belt, and your onions trundle along towards the person on the till, you will notice the conveyor belt always stops before your groceries fall off the end of the belt. This is because your onions break a narrow beam of light. This beam of light hits a photoelectric sensor, which converts the light into an electric current. When the beam of light is broken, the electric current ceases, and the conveyor belt stops.

In the late 19th century, when the photoelectric effect was first discovered, it posed a great mystery. As we have seen in Tesco (or Walmart), the photoelectric effect generates a current when light shines on a metal plate. It was discovered that this current is caused by light knocking electrons from the metal surface. You might imagine that if the brightness of the light was increased, the energy of the emitted electrons would also increase, but this was not what was found. Instead, it was found that the energy of the emitted electrons depended only on the frequency of the incident light. For example, a current was generated when blue light was used (high frequency), but not red light (low frequency). In contrast, if the brightness of the light was increased, the energy of the emitted electrons did not increase – there was just more of them. This posed a great mystery. Once again, it took Einstein to shed some metaphorical light on this particular problem.

Einstein proposed that light was made up of packets of energy, which are now known as photons. Brighter light did not have more energetic photons – it just had more of them. This explained why the energy of the emitted electrons was not dependent on the brightness of the light: an electron was knocked out of the atom by being hit by a single photon, the energy of the photon transferring to the kinetic energy of the electron. The number of photons (the brightness of the light) was irrelevant to the energy of the emitted electron. The only important factor was the energy of the single photon which hit the electron, and that was dependent only on the frequency of the light.

So Einstein revealed that light was "quantized" into particles – individual packets of energy. It was for this discovery – not relativity – that Einstein won his only Nobel Prize in 1922.

It is now believed that all of energy and matter follows this quantum structure and is composed of discrete "chunks" rather than being smoothly continuous. So this first conclusion of quantum theory tells us that the structure of reality is composed of discrete "bits".

On the basis of the two experiments we have considered so far, it appears that light can behave as both a wave and a particle, depending on the experiment that is performed. This might appear strange, but it is all due to the quantum mechanical behaviour of light. It is only by understanding this quantum mechanical behaviour that we can make sense of the behaviour of light.

The third and final experiment we will consider reveals all we will need to know about the logic of quantum mechanical behaviour. As Richard Feynman once said: "The double-slit experiment has in it the heart of quantum mechanics. In reality, it contains the **only** mystery of quantum mechanics."

The double-slit experiment is just Thomas Young's interference experiment (which we considered earlier), but with a slight modification. As we have just seen in the photoelectric effect, it appears that light is composed of particles called photons. In the double-slit experiment, the intensity of the light source is hugely reduced so that only a single photon is emitted at a time. In other words, at any point in time there is only a single photon travelling from the light source, through one of the two slits (we do not know which slit), before hitting the screen behind the board.

We perform this experiment many times, each time when a photon hits the screen it makes a mark. Over time, these marks accumulate. At the end of the experiment, when we investigate the pattern of marks, we find something extremely peculiar. The pattern of marks seems to follow the same pattern of light and dark areas that we found earlier in Young's wave experiment. But in Young's experiment, many billions of photons were being emitted at any moment, with the photons passing through both slits at the same time. The two waves of light from the two slits created the interference pattern. But when there was only one photon travelling in the experiment, how could the interference pattern be formed? To get an interference pattern we would need light to be passing through **both** slits, but a single photon cannot pass through both slits – a single photon cannot interfere with itself.

However, this appears to be what the double-slit experiment is telling us. The result of the experiment only makes sense if the single photon is, indeed, passing through both slits. It is as if the photon can be in more than one place at the same time.

So now we can present the conclusion of the double-slit experiment, and it really does – as Richard Feynman stated – capture the whole essence of quantum mechanics. As we stated at the start of this section, quantum mechanics is really all about measurements, and the double-slit experiment means we can capture the logic of quantum mechanical measurement in just two statements:

1) Before we measure certain properties of a particle (e.g., position, momentum) the particle behaves as though it has **all possible values** for that property.

2) After we have performed the measurement, we find the particle property takes only one of the possible values at random. It is fundamentally impossible to predict which value the property will take.

And that, basically, is all you need to know about quantum mechanics. Armed with these two statements, we can now make sense of the result of the double-slit experiment.

In the double-slit experiment, when the particle hits the screen, we are effectively making a measurement: we are measuring the position property of the photon. Before the photon hits the screen, though, the first statement of quantum mechanical logic states that the photon behaves as though it has all possible values for that measured property: we say the particle is in a *superposition* state. In other words, as the photon leaves the light source, it behaves as though it is in all possible positions. This does, indeed, mean that the photon appears to pass through both slits! Strange as it may seem, this is what quantum logic tells us: before measurement, quantum reality appears to be in a multi-valued superposition. For now, just accept the logic, and be reassured that this quantum logic is applied consistently in all experiments.

The superposition state – in which all the various possibilities are superimposed on the particle – is a crucial and unexplained feature of quantum mechanics. In Chapter Eight I present a speculative explanation for this extraordinary behaviour: a logical rationale for the superposition state.

The second statement of quantum mechanical logic states that after we make the measurement, we find the value for the measured particle property is a single value selected at random. So, in terms of the double-slit experiment, this means that when the photon hits the screen and we measure its position, we find we get a single value – a single mark on the screen. But that appears to imply that the photon only went through one slot, not both. Strange as it may seem, just accept for now that this is what quantum mechanical logic tells us. Before measurement, the particle is in a superposition and behaves as though it passes through both slots, but, after measurement when the photon hits the screen, we find only a single, randomly-selected result.

The quantum casino

The first statement of quantum logic reveals that, before we make a measurement we have to consider a property of a particle to have all possible values. We saw this principle in action in the double-slit experiment when the particle appeared to pass through both slits.

But we have not yet considered the implications of the second statement of quantum logic. That states that, after we make a measurement, the property value is selected at **random** from all possible values. Quantum mechanical behaviour is actually starting to sound remarkably similar to a roulette wheel.

We could imagine a roulette wheel as being a device which determines property values during the process of taking a quantum mechanical measurement. When we considered the photoelectric experiment earlier, we came to the conclusion that reality was composed of discrete "bits": property values could only occupy discrete integer slots (e.g., 1, 2, 3) rather than taking completely continuous values (e.g., 1.342937492 …). So the slots on the roulette wheel represent the only available property values.

When the ball is racing around the wheel, this represents the situation before the measurement is taken. With no idea of what the final result will be, it is as if we must consider the ball as potentially occupying any **and all** of the slots in the wheel (just as a photon in a superposition state is capable of being in more than one place at once, passing through the two slits).

When the ball comes to a rest, it is equivalent to taking a measurement of a property value – the value is the number of the slot in which the ball rests. There is a completely random choice of which value is selected – just like a real roulette wheel. However, the principles behind this randomness are different to that of a normal roulette wheel.

On a normal roulette wheel in a casino it is actually possible to predict (in theory) where the ball will land. If you can measure the precise speed at which the wheel spins, and you can measure the precise speed that the croupier throws the ball, then, by using our knowledge of classical physics and Newton's laws of motion it is theoretically possible to work out where the ball will finish. It is not easy, but it is possible.

The quantum casino works differently. In the quantum casino there is no way of calculating where the ball will land. In a normal casino, there would be a bouncer preventing you from making your measurements. You would be prevented from measuring the speed of the wheel. You would be prevented from measuring the speed of the ball. You would be prevented from seeing that information. Similarly, in the quantum world when we are measuring a property of a particle we are fundamentally prohibited from digging deeper to analyse the quantum mechanism to try to predict the result. That is a fundamental limit on physics. We just have to accept that the result of the measurement we get will be random. We can never do any better. This is why we have to consider the quantum behaviour of Nature as being **fundamentally** random – because no deeper layer can ever be accessible to our analysis.

The only information we can extract from a system has to be obtained via measurement. In the double-slit experiment, the only information we can obtain is when we measure the photon position when the photon hits the screen. We have no more information about the earlier trajectory of the photon because we have performed no other, earlier measurements. For this reason, we have no deeper information about the photon's path: without taking measurements, that information is fundamentally hidden from us. Comparing with the roulette wheel, we can obtain no deeper information about the speed of the wheel or the speed of the ball in order to predict the result.

But, if we decide to take an earlier measurement to help our prediction, to remove the randomness, we fundamentally change the experiment. In the double-slit experiment, if we put any form of sensor on each of the slits to determine which slit the photon has passed through, we do indeed detect the photon passing through just one slit – not both.

But this has the effect of destroying the interference pattern on the screen – we only see the interference pattern when the photon passes through both slits.

So quantum randomness is not a result of our ignorance of the situation. We cannot determine the behaviour of the system by attempting to reduce our ignorance by taking more measurements. If we take more measurements, we irretrievably alter the experiment.

This random behaviour reveals something very fundamental and remarkable about quantum mechanics: quantum mechanics only tells us the probabilities of a measurement outcome taking a particular value – quantum mechanics can never make an exact prediction about which value will result. This is because quantum mechanics is a statistical theory: it cannot accurately predict the outcome of a single measurement, it can only give the probabilities of outcomes when we make a series of measurements. You might think that is a major shortcoming of quantum mechanics, but actually it can be a very powerful tool to know the probability of an outcome, especially when you are dealing with millions of random events (as is the case with millions of particles). In that case, with millions of measurements being averaged, the randomness appears to vanish. No casino knows the outcome of each spin of the roulette wheel, but casinos know that probability is on their side. And you never see a poor casino!

Schrödinger's fridge

Clearly, our discussion of the double-slit experiment has raised some questions about the nature of quantum reality before observation. This also raises the more general question about what it is possible to say about everyday, human-scale reality (i.e., macroscopic reality) before we observe it. These questions can be summed-up by the example of the light inside the refrigerator.

Imagine you have a refrigerator whose door is closed. Your task is to find whether or not the light is on inside the fridge. In order to find out, you open the door of the fridge wide, and you can then tell that, yes, the light is definitely on inside the fridge.

However, I am sure you can see the flaw in this observation. The very act of observation (opening the fridge door) modified the object under observation (the light is turned on). It would appear there is no way of making the observation without modifying the object under observation.

Of course, at macroscopic scales it is always possible to modify the experimental set-up to obtain more information without significantly modifying the object under observation. For example, an additional light sensor could be placed inside the fridge, and this could communicate the state of illumination inside the fridge via a wireless connection. However, at the quantum level, this is simply not an option. As we discussed in the double-slit experiment, if we put any form of sensor on each of the slits to determine which slit the photon has passed through, we irretrievably modify the experiment. By adding a sensor to determine which slit the photon passed through, this has the effect of destroying the interference pattern on the screen.

The Copenhagen interpretation

If our fridge was a "quantum fridge" then we would be fundamentally prohibited from determining if the light was on or off before we open the door of the fridge. In that case, what can we possibly say about the reality of the fridge light before we open the door? It would seem that we are prevented from finding out what is going on before we open the door.

In quantum mechanics, this limitation opens the possibility to create any number of different interpretations of what is happening to reality "behind the scenes" before we make a measurement. As long as your interpretation makes accurate predictions about the result **after** measurement, then you are free to make whatever interpretation you like about what happens **before** measurement. No one can claim that their interpretation is superior to your interpretation. As a result, there are many different interpretations of quantum mechanics.

The first interpretation of quantum mechanics was proposed by the discoverers of quantum mechanics in Copenhagen in the 1920s. This so-called Copenhagen interpretation was for many decades effectively the official interpretation, and was taught in schools and colleges as though the matter was settled, when, in fact, this is far from the case. As Murray Gell-Mann put it: "Niels Bohr brainwashed a whole generation of physicists into believing that the problem had been solved."

The Copenhagen interpretation is only interested with the results of experiments, i.e., the result we get when we make a measurement. So, according to the interpretation, we should simply not talk of the reality of the system before we make our measurement – it would be unscientific, the subject of philosophy, to consider that.

In taking this approach the Copenhagen interpretation neatly sidesteps any philosophical questions such as "Does the particle exist before we observe it?" But there is a price to be paid for this apparently neat tying together of the loose ends. In fact, there are a couple of unsatisfactory implications of this interpretation.

Firstly, the interpretation does not state what, exactly, constitutes an observation. This seems like a major shortcoming, especially when the act of observation is such an important factor in the Copenhagen interpretation. As we have seen in the double-slit experiment, before observation, a particle is in a superposition state, appearing to have all possible property values. However, after observation, the particle appears to have only a single property value. What possible process could cause this sudden "quantum jump"? This is called the *quantum measurement problem*, and its resolution remains a major challenge in quantum mechanics.

What does it take to produce an "observation"? Can a machine take an observation, or does it require a human being?

In his 1932 book *The Mathematical Foundations of Quantum Mechanics*, the great mathematician John von Neumann suggested that the consciousness of the experimenter is the last link in the chain which brings objects into reality during a measurement process: "There exist external observers which cannot be treated within quantum mechanics, namely human (and perhaps animal) minds, which perform measurements on the brain causing [quantum observation]."

So does it really take a conscious human being, performing an observation, to generate reality? Surely not. For example, a radioactive uranium nucleus buried in rock on a distant planet will decay to emit an alpha particle. It does not matter if a human observer looks at the rock or not. Einstein, who never accepted the implications of the

Copenhagen interpretation, summed up the absurdity of the situation when he said: "Do you really think the moon isn't there if you aren't looking at it?"

As Carver Mead has said:

> *That is probably the biggest misconception that has come out of the Copenhagen view. The idea that the (human) observation of some event makes it somehow more 'real' became entrenched in the philosophy of quantum mechanics. Even the slightest reflection will show how silly it is. An observer is an assembly of atoms. What is different about the observer's atoms from those of any other object? What if the data are taken by computer? Do the events not happen until the scientist gets home from vacation and looks at the printout? It is ludicrous!*

In his statement about the unimportance of human consciousness in the act of observation, Carver Mead was probably influenced by the famous thought experiment of Schrödinger's cat. Erwin Schrödinger presented the tale of his cat in 1935 to reveal the shortcomings of the Copenhagen interpretation (which was dominant at the time), and, specifically, to reveal the absurdity of the idea that only human consciousness can produce an observation.

Schrödinger's cat is placed inside a sealed box. In the box with the cat is a sealed glass jar of poison. There is also a small piece of radioactive material. If the piece of radioactive material decays (the decay being dependent on quantum randomness), the poison is released and the cat dies. If the radioactive material does not decay, the cat lives.

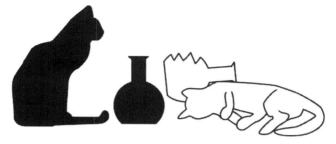

According to the "human consciousness" interpretation, the reality of the radioactive decay is only determined when it is examined by a human being. This would mean that, until the box was opened and the radioactive material is examined by a human, we must consider the material to be in a superposition state of both "decayed" and "not decayed". We must therefore treat the cat as being in a state of both "dead" and "not dead"!

From this *reductio ad absurdum* argument, we have to conclude that human consciousness is not the only mechanism capable of making an "observation" and reducing a quantum superposition to a single value.

So how can we define an "observation" in a more general sense? An excellent definition can be found in the book *Quantum Enigma* by Bruce Rosenblum and Fred Kuttner: "Whenever any property of a microscopic object affects a macroscopic object, that property is 'observed' and becomes a physical reality." For example, when a microscopic photon hits the macroscopic screen in the double-slit experiment, then that will reduce the quantum superposition state of the photon to a single value (the single mark it leaves on the screen). This explains why we do not see bizarre quantum superpositions – such as a cat being both alive and dead at the same time – in the human-scale, macroscopic world. So as long as there is a macroscopic effect from a quantum entity, that object can be considered to be "observed" or "measured" – with no need for a conscious human observer.

But, note we have said nothing about the details of the physical process of observation. What actually occurs at the quantum level when an "observation" is made? For a detailed discussion of the act of observation, we must wait for the next chapter.

The Many Worlds interpretation

In a recent survey, the Copenhagen interpretation remains the most popular interpretation of quantum mechanics among physicists. However, in second place (and rising fast) we find the Many Worlds interpretation (MWI). The MWI was proposed by Hugh Everett in 1957, and might be regarded as the "parallel universes" interpretation. If you remember from earlier in this book, I am certainly not an advocate of theories based on parallel universes, and I do not believe they should be given the same credence as conventional theories. In this section, I will describe the MWI, and in the next chapter we will see why I believe it is wrong.

The MWI attempts to solve the main two problems associated with the quantum measurement problem:

1) Before observation, a particle is in a superposition state of all possible values. During measurement, what causes the reduction of this state to a single value?

2) Before observation, we have many values. After observation, we only have a single real value. So what happened to all the other values that never became "real"?

According to the MWI, when we make an observation – for example, when the particle hits the screen in the double-slit experiment – the universe splits into a series of parallel universes. In each parallel universe, each possible value of the quantum superposition becomes real. For example, in the double-slit experiment, you will remember that the single photon appears to pass through both of the two slits. So when the particle hits the screen, the MWI proposes that the universe splits into two parallel universes. In one universe, the particle passes through one of the slits, and in the other universe, the particle passes through the other slit. As another example, in the case of Schrödinger's cat, when we open the box the universe splits into two: the cat is alive in one universe, and dead in the other.

This must necessarily mean that the entire universe splits – including you! Every time any measurement is made, you must split to make a duplicate of yourself, and that duplicate resides in a different parallel universe.

This simple (though extravagant) explanation appears to solve the two problems associated with the quantum measurement problem given earlier. Firstly, it explains how a quantum superposition is reduced to a single value. The answer, according to the MWI, is that there is no reduction, no non-linear quantum "jump". All possible values of the quantum superposition become real after measurement, we just happen to inhabit one of those universes so it appears the particle only goes through one slot. However, we will be returning to consider the validity of this "linear" MWI claim in the next chapter.

The MWI also explains the second problem associated with the quantum measurement problem: what happens to all the values of the superposition which do not become real? The answer, according to the MWI, is clearly that all values in the quantum superposition become real – in different parallel universes.

Throughout this chapter on quantum mechanics I have emphasized the philosophical questions about quantum reality before observation. Hence, the title of the chapter is "quantum reality" rather than "quantum mechanics". It is quite extraordinary how the so-called theories of everything – string theory and loop quantum gravity – say nothing at all about this crucial concept of quantum reality nor the measurement problem. This is all the more astonishing as it is my firm belief that the principle behind quantum reality plays the vital role in providing the link between relativity and quantum mechanics.

7

OBSERVING THE OBSERVER

In the discussion in the last chapter on quantum reality, we were left with a big question mark over the role of the observer. This is a major omission, as the observer plays such a central role in quantum mechanics. In order to resolve the quantum measurement problem, it seems essential to define the nature of the observer, and to determine what, exactly, occurs during an "observation".

In the Copenhagen interpretation, the role of the observer is to bring an object into reality. Before observation, the object appears to be in a peculiar, multi-valued superposition state, and it is only after observation that we find a single, well-defined reality. The role of the observer is clearly crucial in this case. As for defining the nature of the observer, as we discussed, it was believed at one point that only a conscious human observer was capable of performing an observation, though I think we can discount this theory.

In the Many Worlds interpretation (MWI), the role of observation is again crucial as it is the process of observation which causes the universe to split. However, the MWI claims to avoid the need for any non-linear quantum "jumping" during the observation process, as is required by the Copenhagen interpretation (we will be examining this claim in detail later). According to the MWI, not only does the object under observation split into a different universe, but the observer also splits. Considering the example of Schrödinger's cat, according to the MWI, there would be one universe with a dead cat (which contains an observer who observes the cat as dead), and another universe with a living cat (which contains an observer who observes the cat as being alive). It would appear no non-linear quantum "jumping" is required. However, the price to be paid for this elegant solution is quite extravagant: there are a lot of dead cats (and parallel universes) being produced.

However, it is the belief of this book that the proponents of both the Copenhagen interpretation and the MWI did not fully appreciate the implications of a relative universe. The paradoxes which are associated with the quantum measurement problem only arise because a model of the universe is used which is not fully relational. By using a fully relational model of the universe, all the confusion about the quantum measurement problem can be eliminated – no need for the "quantum jumping" of Copenhagen, or the parallel universes of the MWI. Everything can be explained quite rationally and logically, in a single, relational universe.

Amazingly, as in every example in this book, the solution to the quantum measurement problem can be found in the principle that Nature has no access to absolutes.

Lost in space

Our first task is to get the definition of the "observer" and "observation" onto a firmer footing. Let us return to consider the example of the two astronauts, floating past each other in empty space. We initially considered this example in our discussion of relativity in Chapter Three. Imagine you are an astronaut, floating in the emptiness of space. On your right-hand side, you **observe** (note the emphasis) a fellow astronaut floating past you, travelling at a constant velocity. Your friend passes by you, before vanishing into the darkness.

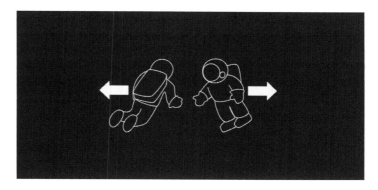

Because Nature is unable to access any absolute axes of position, there is a perfect symmetry in this scenario. No astronaut can claim to be preferred. The experience of both astronauts is identical. Nature cannot assign a stationary position to one astronaut, and a moving position to the other – neither astronaut has a greater claim to be stationary. As a result, both astronauts feel stationary throughout.

So, here is an interesting thought. If, as we described, you **observed** the other astronaut throughout the scenario, then, due to the symmetry of the situation, it is obviously equally valid to state that the other astronaut observed you in precisely the same way throughout. Just as Nature could not assign one astronaut to be stationary and one moving, Nature clearly cannot assign one astronaut to be the "observer" and the other to be the "object under observation".

We are starting to see that – as far as Nature is concerned – there is simply no fundamental distinction between "observing" and "being observed". The distinction is purely a human invention.

You could replace the other astronaut with an unmanned spaceship floating past at constant velocity, and argue that the astronaut is capable of "observing" the spaceship, but it is impossible for the spaceship to "observe" the astronaut as the spaceship is not a conscious entity. However, in trying to preserve the distinction between "observing" and "being observed" in this way, you are having to reintroduce the notion of "conscious observation" which we considered and refuted in the last chapter on quantum mechanics. As far as Nature is concerned, it is just as valid for an inanimate object to observe a conscious object, as it is for a conscious object to observe an inanimate object.

Fundamental observation

To understand how Nature views the concept of "observation" we have to examine the lowest scale, the level of particles. We find that even at this fundamental level, Nature makes no distinction between "observing" and "being observed".

To show this, let us imagine we want to observe an electron to find out its current position. Essentially, this means we want to measure its position property. The obvious way to observe anything is to look at it. This works perfectly well for everyday macroscopic objects (for example, when you lose your keys), but does the principle still hold at the fundamental level of particles? Can you look at an electron, for example, to find where it is?

Well, it turns out that you can. The simplest way to detect the location of an electron is to shine a light on it. But, as we discovered in the previous chapter, at the fundamental level we find that light behaves as if it was made out of small packets of energy called *photons*. When we hit the electron with a photon, the electron absorbs the photon (temporarily raising itself to a higher energy level) before emitting the photon again. So by attempting to observe the electron by using the photon, we inevitably disturb the object we want to observe.

At the fundamental level, all Nature consists of is interactions – collisions – between elementary particles. This can be seen in the following Feynman diagram which shows electromagnetic repulsion between two electrons. The two electrons come in from the left of the diagram, they exchange a photon between themselves (represented by the wavy line), and this has the effect of throwing them both off course, just like billiard balls:

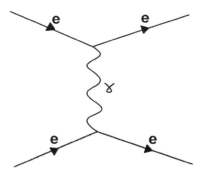

We can simplify this diagram by hiding the nature of the collision (i.e., we are not interested in the behaviour of the photon), and just showing the deflection of the electrons when they collide:

This diagram reveals the symmetry of the interaction. Once again, this is caused by Nature being unable to access any absolute axes of position. No particle can claim to be preferred. No particle can claim to be the "observer" with the other particle "being observed". Both particles are on an equal footing.

When we attempt to "observe" the electron using the photon, we inevitably disturb the electron – the supposed object "being observed". So the electron (the object "being observed") alters the course of the photon (the "observer") as we intended. But the photon ("the observer") also alters the course of the electron (the object "being observed") in a way which we did not intend. As a result of the symmetry of the situation, it is just as valid to say that the electron "observed" the photon as it is to say that the photon "observed" the electron!

As we discovered earlier in our discussion of the two floating astronauts, we again see that – at the fundamental level – as far as Nature is concerned, there is simply no fundamental distinction between "observing" and "being observed". The distinction is purely a human invention. Whenever you observe an object, it is just as valid to say that the object is observing you!

It is truly a case of the object under observation "observing the observer":

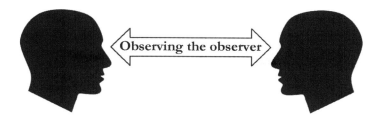

This symmetry reminds us of another symmetrical law, Newton's third law of motion: "Every action has an equal and opposite reaction". Essentially, if I press on a stone with my finger, the stone also presses onto my finger with a precisely equal and opposite force. Forces are interactions between two bodies, and those forces are always equal and directed in the opposite directions. Once again, this symmetry is caused by Nature's fundamental inability to distinguish between two situations: as far as Nature is concerned, your finger pressing on the stone is essentially equivalent to the stone pressing on your finger. Nature views your finger and the stone as just groups of atoms – it has no access to any absolute scale by which it could distinguish the two objects.

We are reminded of the two astronauts floating past each other in space: the experiences of both astronauts is identical. The observation made of the second astronaut by the first astronaut is equivalent to the observation made of the first astronaut by the second astronaut. Let us turn that into a general principle:

The observation of an object by an observer results in an equivalent observation of the observer by the observed object.

Maybe we should call this the "principle of the two astronauts"!

Of particular interest to quantum mechanics (in which it is impossible to measure a particle without modifying it):

The physical action on the observed object by the observer will be equal and opposite to the physical action on the observer by the observed object (this is just a restatement of Newton's third law of motion: "Every action has an equal and opposite reaction").

Armed with these two principles, we now have a much greater understanding of what is involved during an "observation", and we are now in a position to reveal what really happens during a quantum mechanical observation. We can at last attempt to find a solution to the quantum measurement problem.

Environmental decoherence

Throughout this book we have emphasized that there is only one universe, and the universe should be treated as a single, connected object. We will now see that this provides the vital clue to solving the quantum measurement problem.

In order to understand the process of observation at the quantum level, and to solve the quantum measurement problem, it is essential to treat objects as being connected and interacting, rather than being isolated and independent. The key to solving the quantum measurement problem is to treat the object under observation and the measurement apparatus as a single, connected, interacting system. We not only have to consider the effect of the object under observation on the measuring apparatus (e.g., moving a dial on a meter), but we also have to consider the effect of the measuring apparatus on the object under observation.

We all have a natural, human bias to imagine objects under observation as isolated and independent of any measuring apparatus. This is because, at the human-scale, macroscopic level, when we measure an object we do not notice any significant change in the object under observation. For example, if we are measuring the air pressure in a tyre, we would use a pressure gauge – which inevitably results in a small quantity of air leaking out of the tyre. Hence, by measuring the pressure in the tyre, we are inevitably altering the object under observation (we are reducing the pressure in the tyre). However, the reduction in pressure would not be noticeable when compared with the overall pressure of the tyre – it would be an insignificant fraction. So at the macroscopic level, we simply do not have to worry about how our observations and measurements affect the object under observation.

This is not the case at the fundamental, quantum level. The Heisenberg uncertainty principle showed we cannot observe or measure an object without fundamentally altering it. At the quantum level, the effect of the measuring apparatus on the object under observation becomes unavoidable.

It was not until the 1970s that the effect of the role of the measurement apparatus on the object under observation was considered in detail. We have seen that a measurement apparatus is capable of reducing a quantum superposition state to reveal a single measured value, but there is nothing particularly special about measurement equipment which gives it this ability. A simple screen, for example, is capable of revealing the position of a photon in the double-slit experiment. As we quoted from the book *Quantum Enigma* earlier: "Whenever any property of a microscopic object affects a macroscopic object, that property is 'observed' and becomes a physical reality". In the 1970s, it was revealed that the general environment around a particle was capable of reducing the quantum superposition of a particle. The process by which this occurred was called *environmental decoherence*.

Essentially, the general environment (air molecules, photons, dust) performs the role of the measurement apparatus in that it betrays the position (or other property) of the particle under observation. It is the effect of the environment on the particle under observation which explains why we do not see quantum superpositions at the human scale.

So how does quantum decoherence work? For example, if we are measuring some property of a particle using a meter with a dial, how would the particle (in a multi-valued quantum superposition state) be reduced to a single-valued state? Where would the other interference states go?

The discussion in the previous section has given us a good basis for understanding what actually happens in an "observation". The key element was seen to be the symmetry of the observation. We have seen that Nature makes no distinction between the observer and the object under observation. We have seen that any physical effect the object under investigation might have on the observing apparatus must be balanced by a symmetrical effect which the measurement apparatus has on the object under observation.

Considering the example of a single particle being measured by a meter, we are trying to measure the value of some property of a single particle – using a meter which is composed of billions of billions of particles. Straight away, we can see a clear asymmetry in the set-up – this is a clue as to why we get the result we see. However, let us ignore the discrepancy between observer and observed for the moment, and consider what happens to a single particle.

As we discussed in the previous section, when we attempt to measure a particle, it does not matter how big your measurement apparatus is, all measurements at the fundamental level have to be considered as interactions between individual particles.

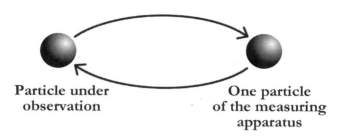

Particle under observation

One particle of the measuring apparatus

As far as Nature is concerned, at this level there is no distinction between the observer and the particle under observation. There is a perfect symmetry. There **has** to be a perfect symmetry. We have reached this conclusion by building-up from fundamental principles: this is caused by Nature being unable to access any absolute axes of position. No particle can claim to be preferred.

So now let us expand our model to consider the entire measurement apparatus. Clearly, this consists of vastly more particles, but the basic principle remains the same: all measurement must consist of symmetrical interactions between individual particles. The diagram on the next page shows the first half of this symmetrical interaction – the effect of the particle under observation on each particle in the measurement apparatus:

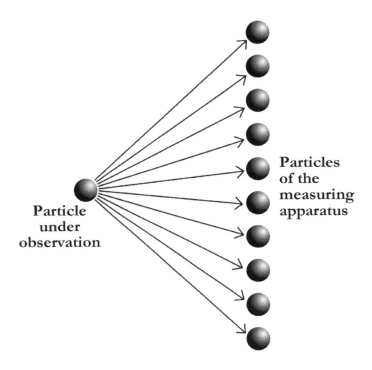

Clearly, the effect of the single particle on each particle of the measuring apparatus is going to be small – those are small arrows I have drawn on the diagram above. The effect of the single particle is dissipated through all the billions of atoms in the measuring device and becomes effectively unnoticeable at the macroscopic level.

However, the interactions between particles has to be symmetrical. Each of those little arrows drawn on the diagram above has to produce a symmetrical reaction from each particle of the measuring apparatus back onto the single particle being measured. It is this combined reaction which is devastating for the single particle being measured:

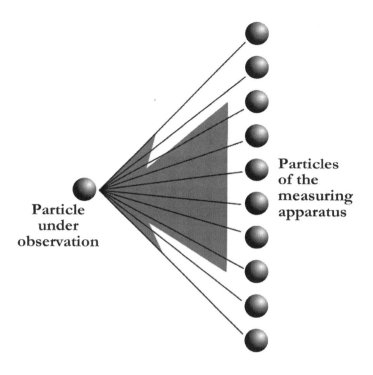

The asymmetry of this arrangement shows that the impact of the measuring apparatus on the particle under observation is hugely larger and more significant than the impact of the particle under observation on any single particle of the measuring apparatus. The result of this huge impact is to reduce the superposition state of the particle under observation to a single, measured value.

The details of how this reduction of state is achieved is not so important (it is an active research topic, and there are various theories) – it is the asymmetry of the arrangement which is the important and decisive factor. However, it appears it is the randomness of the environment which destroys the regularity of the peaks and troughs (the *phase*) of the particle's wavelike nature. If we consider the double-slit experiment, for example, we only see the interference pattern on the screen (representing the particle's superposition state) if the regular *coherent* phases of the particle's waves are not disturbed. Any randomization of the phases causes the particle to *decohere* – the superposition interference terms dissipate unnoticed into the general environment. As Brian Greene said in his book *The Fabric of the Cosmos*: "Decoherence forces much of the weirdness of quantum physics to 'leak' from large objects since, bit by bit, the quantum weirdness is carried away by the innumerable impinging particles from the environment." For all intents and purposes, the interference effects have completely disappeared. The particle is reduced from a superposition state to a single-valued state.[13]

Environmental decoherence provides a simple solution to the paradox of Schrödinger's cat. Long before the box is opened by the human observer, the environment inside the box (e.g., air molecules) has interacted with the radioactive material and "observed" it to discover if it has decayed or not. The state of the cat – either alive or dead, but not both – is set long before the box is opened.

Environmental decoherence also avoids the need for any non-linear quantum "jump" during the observation process. Previously, it had been believed that there was an unexplained jump between the particle's multi-valued superposition state and the single-valued final measurement. Decoherence avoids this jump by revealing that all the interactions between the observed particle and the measuring apparatus are perfectly linear – no jumping required. Instead, the interference terms disappear due to the progressive

---

[13] For technical details of environmental decoherence, see the page on my website: http://www.whatisreality.co.uk/reality_decoherence.asp

influence of billions of particles as the particle under observation passes through the measuring apparatus. So there is a progressive filtering (of the interference terms) and amplification (of the eventual measurement). Niels Bohr referred to an "irreversible act of amplification". In this respect, the measuring device behaves very much like a radio receiver, picking out a signal on a single frequency and amplifying it, while simultaneously reducing the interference signals on all the other frequencies.

Though it appears that there is an instantaneous non-linear "jump" to particular defined state, it emerges that the process is, in fact, linear, and only appears like a sudden jump because of the sheer speed of the decoherence process. Decoherence is believed to happen in the region of 10^{-27} seconds, and is therefore one of the fastest processes known to science – but it is not instantaneous.

Experimental evidence for environmental decoherence is now emerging. In 1996, a team at the National Institute of Standards and Technology (NIST) in Boulder, Colorado isolated an atom from the environment by trapping it in an electromagnetic force field. Then, when they hit it with a laser beam they managed to force it into being in two places at once, separated by about a thousand times the diameter of an atom. Other teams based at the State University of New York and the Technical University of Delft managed to force an electric current to flow around a ring in opposite directions at the same time![14]

So we now have the basis for an explanation for the quantum measurement problem. We can see why the measurement apparatus appears relatively unchanged, while the superposition state of the particle under observation is

_____

[14] *New life for Schrödinger's cat* by Tony Leggett:
http://www.whatisreality.co.uk/reality_schrodingers_cat.asp

reduced. The key concept to remember is that we cannot treat particles as being isolated from the rest of the universe. We cannot consider the particle in isolation from the measurement apparatus. We have to consider the universe as a single, connected object.

Irreversibility

One of the features of quantum "jumping" is that it is irreversible. We only ever see quantum jumps in the forward time direction, from a superposition state to a well-defined state – never vice versa. This remains an unexplained mystery for some interpretations of quantum mechanics – such as the mystical Many Worlds interpretation – but can be simply explained in environmental decoherence.

There is nothing mystical about environmental decoherence – it is just a physical process just like any other physical process. Hence, it comes under the remit of the second law of thermodynamics (remember back to our discussion of the arrow of time in Chapter Five). Indeed, the irreversibility of quantum jumping in the forward time direction – just like the irreversibility of a wine glass breaking in the forward time direction – is the greatest clue that increasing entropy is the underlying mechanism behind quantum jumping. Just like a wine glass breaking, or a car rusting, the entropy of a closed system will always increase with time. Likewise, when an isolated particle encounters a randomised environment, the effect of the environment will be to destroy the ordered coherent phases of the particle's wavelike nature. The regularity dissipates into the environment and can never be reformed – just like a broken egg.

Imagine you throw a rock in the sea off the coast of the United Kingdom. After the initial big splash, the ripples dissipate and apparently disappear. But of course, they have not really disappeared. The ripples have decreased in size as they have mixed and interfered with other waves, but they have not disappeared. Two weeks later, on the rocky shore of Tierra del Fuego off the Argentinean coast, one of the small waves washing to shore is maybe an imperceptible fraction of one micron higher because of that rock you threw.

So the ripples (interference terms) do not actually disappear. They dissipate into the environment and become effectively undetectable. It is certainly not possible to associate the microscopic change in the height of the wave in Tierra del Fuego with the rock you threw – there have been so many interactions with other waves along the way. In this sense, the process of decoherence is irreversible. We can't reverse the process to regenerate the initial interference components – they are gone for good. And even the "little ripple" echoes of the interference effects have become imperceptible due to interactions with the environment. Then, **for all intents and purposes**, the interference effects have completely disappeared.

Why Many Worlds is wrong

The early interpretations of quantum mechanics – such as the Copenhagen interpretation and the Many Worlds interpretations – pre-dated the discovery of environmental decoherence. Both interpretations were aware of the importance of the observer, but neither managed to accurately model the physical process of observation.

In the Copenhagen interpretation, it was suggested that the effect of the observer was to somehow make the object under observation "real", and we simply should not ask questions about the state of reality before observation, or what occurs during the process of observation.

In the Many Worlds interpretation (MWI), the modelling of the observer was more subtle, with an attempt being made to improve accuracy. It was realised that the object under observation and the measuring apparatus had to be treated as a single, composite system. It was then realised that, if all the multi-valued components of the superposition state were to become real after observation, the observer would have to split into many different versions as well, with each observer viewing a different component of the superposition.

However, the MWI fails in that it models the fundamental interactions in a completely asymmetric manner. As we have said throughout this book, Nature is fundamentally based on symmetries, an inevitable consequence of the limitations imposed on Nature. But we do not find this symmetry in the MWI. In the environmental decoherence model, the effect of the single particle on the measurement apparatus is effectively unnoticeable as the effect dissipates through the billions of particles in the environment. However, in the MWI, the effect of the single particle on the measurement apparatus is disproportionately huge – it is far from unnoticeable. In fact, it is enough to generate several copies of the measurement apparatus. In the MWI, the effect of the single particle is considered equal to the power of the entire measurement apparatus – which is composed of billions of billions of particles!

As we calculated from fundamental principles, interactions between particles should be symmetrical. This is the unquestionable, logical conclusion. While this is the case for the environmental decoherence model, it is anything but true for the MWI. The MWI therefore breaks fundamental, logical principles.

But what about the claim that the MWI is a linear solution, which avoids any problematic quantum jumping? Let us examine this claim in more detail.

The great puzzle of the quantum measurement problem is that, after measurement, all but one component of the quantum superposition appears to vanish as the object under observation "jumps" to a single, well-defined state. In the example of Schrödinger's cat, the cat in the dead/alive superposition state only appears as either dead or alive after observation. So where did the other state of the cat go?

The MWI attempts to solve this paradox by considering the measurement apparatus and the cat as part of a single entangled system. In order to illustrate this, let us introduce a cat detection meter which observes the cat as being either alive or dead (this represents the role of the human observer opening the box to observe the state of the cat). The following diagram shows the cat detection meter in its initial state — before an observation has been made:

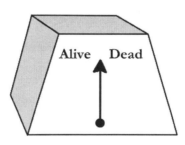

Now, according to the MWI, when a measurement is made, the dual-nature of the cat superposition ensures that there are two copies of the meter registering the two possible states of the cat:

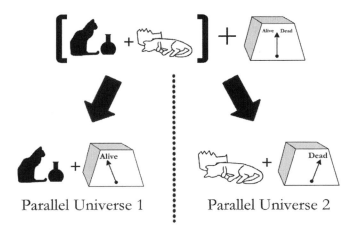

Parallel Universe 1   Parallel Universe 2

Advocates of the MWI would say this represents linear evolution: both states of the cat before observation – alive and dead – appear in reality after observation.

However, this claim of linearity is misleading. This is because the cat detection meter (the measurement apparatus) was **not** in a superposition state before the measurement – it was in a single, well-defined state. Try firing the meter at a double slit, for example, and it definitely will **not** produce an interference pattern! Likewise, the corresponding human observer was not in a superposition state: there were definitely **not** two copies of the human observer. In the MWI, however, we find that – after the observation – the measurement apparatus has jumped to be part of an entangled superposition: there are effectively two versions of the measurement apparatus after the observation.

In graphical form, here we see the problematic quantum jump in the state of the object under observation in the conventional quantum measurement problem:

And here we see a corresponding quantum jump in the measurement apparatus in the MWI:

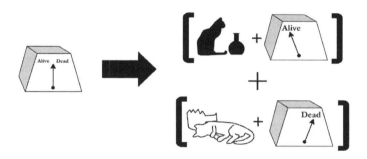

So in the MWI, the problematic quantum jump is just transferred from the object under observation to the measurement apparatus. The MWI shuffles the two outcomes off into two different "parallel universes" like a criminal attempting to hide the evidence of a crime, but it is clear that, as far as linearity is concerned, the MWI is no improvement over the existing quantum measurement problem – despite the claims of its advocates. The same cannot be said for environmental decoherence which remains perfectly linear throughout the observation process.

Throughout this book I have stressed that there is no evidence for parallel universes, nor is there any convincing rationale to adopt these theories. Only by treating the universe as a single, connected object are we able to uncover the true secrets of the workings of Nature – such as environmental decoherence.

In the next chapter we will see how the connectedness of the single universe provides an explanation for quantum mechanical behaviour.

8

THE QUANTUM RATIONALE

This is a book about reality. It is about *physical* reality. By limiting the term "reality" to mean physical reality we are immediately rejecting any notion that dreams, or ideas, have any form of reality – at least by our definition. Reality – in our book – refers to physical objects we can touch and hold. This means the material of reality is, essentially, matter.

Our definition of reality is also about *objective* reality, objects which everyone agrees are "real". This is to avoid any possible confusion with *solipsism* – the idea you are the only real consciousness in existence, and everyone else (and everything else) is just a figment of your imagination, a fictitious dream world – just like season eight of *Dallas*.

Though it can never be possible to disprove that you are living in a solipsistic world, we have to assume that we are not or else we have to accept that nothing is real and there is no point analyzing the universe around us. We have to continue our discussion on the basis that everyone's consciousness is equal, and that there is no "master consciousness" in which everything else exists only as a dream. If everyone agrees that an object is real, then we accept that it truly is real and is not just a figment of our imagination.

But this leaves the challenging question: what do we actually mean when we say something is "real"? We have an intuitive notion of what the word means which is satisfactory for our purposes, and we all use the term in our everyday vocabulary without thinking too deeply about the meaning of the word "real". But what, exactly, does it mean? What property does a "real" object have which determines its reality? How does something become real? Can an object become unreal? These are questions to which we cannot provide a satisfactory answer because we simply do not appear to have an answer to the question of what, precisely, does it mean to be "real" in the first place. We have to be careful to avoid using any linguistic terms – which might well be in common usage – without making sure those terms have clear definitions. If we are going to be serious about our quest to uncover the nature of reality, we had better make sure we know just what we mean by "reality".

David Deutsch in his book *The Fabric of Reality* suggested that real objects should "kick back". For example, if we kick a rock with our foot, the rock makes sure it "kicks back" and stubs our toes. This suggests that if something is outside of our consciousness, and is able to interact with us, then it forms an independent, objective reality. I would agree with David Deutsch, and this is a good example of how we decide an object is real. But I believe his explanation of the example does not fully explain the underlying principle by which we define real objects – how we truly know that something is real when we kick it.

The key is that our reality is relative, not absolute. We define real objects in terms of other objects which we already consider to be real, or tangible. For example: "I know the apple is real because I can hold it in my hand". The apple's reality is defined in terms of the presumed reality of the hand. That is the best definition of physical reality we can ever possess, a rather circular definition.

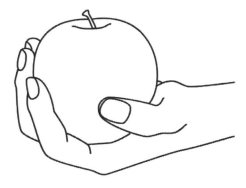

We can see that the whole of reality is composed of a self-supported structure of relationships, objects which are assumed to be real because they interact with other objects which are presumed to be real. There is an interconnectedness to reality, a unity. While this might seem like a very circular state of affairs, it is inevitable in a relative universe.

Reality is relative

In the previous chapter, we stressed the importance of the connectedness of the universe, and this provided us with a proposed solution to the quantum measurement problem in the form of environmental decoherence. This connectedness is, once again, an inevitable consequence of there being nothing outside the universe, and the resultant lack of absolutes available to Nature. This leads us to an extremely important conclusion: if there is nothing outside the universe, and no absolutes, then:

The properties of any particle – or any object – are determined by all the other objects in the universe.

This is an incredibly important principle, but it seems to be not generally realised. All properties of all objects are only defined relative to the other objects of the universe.

As we have said many times before, Nature has no access to absolutes, and so would be unable to assign absolute property values to the object – Nature has no access to an absolute standard of measurement. The only possible values Nature could assign would be relative values, relative to other objects in the universe.

But what if Nature is unable to assign these relative property values? Imagine an object which is the only thing in existence – there is nothing else in the universe. What would it mean to assign property values to this isolated object? Surely, with no other objects in the universe, such property values would be meaningless. What is the point in assigning a certain mass to the object if it is unable to interact with other objects? What is the point in assigning a velocity value to the object if there is no other object by which that relative velocity could be measured? On their own, property values are meaningless. Amazingly, we see that properties do not describe the isolated object itself, **properties describe the relationships between objects** – they describe the effect we will see when that object interacts with other objects. As the great Niels Bohr said: "Isolated material particles are abstractions, their properties being definable and observable only through their interactions with other systems."

This principle arises because the universe is a relative structure – all objects in the universe must be defined relative to all the other objects in the universe. With no absolutes available to Nature, there is simply no other way by which objects can be defined. Even though this seems like an extraordinarily important principle, I hardly ever see it mentioned. Lee Smolin is one of the few who appears to realise: "The fundamental properties of the elementary entities consist entirely in relationships between those elementary entities." And again: "A relational view requires that the properties of any one particle are determined self-consistently by the whole universe."[15]

Back in Chapter Two we considered the importance of units in our measurements. Units provide the essential relational link between the unknown object being measured and the rest of the known world. In the absence of any external absolute scale, the only way we can define an object's properties is by measuring it with equipment already located inside the universe, i.e., the object's properties are defined by its relationships with other objects inside the universe. The object's properties simply **cannot** be defined in any other way. Again, this shows that meaningful properties are not inherent in an object, but the meaning is determined by the other objects in the universe. For example, the length property of a plank of wood arises from its relationship with the measuring tape. Giving property values to isolated objects is meaningless – that would be equivalent to omitting units in your measurements and just saying "This goes up to 11".

A perfect example of this principle is provided by the recent discovery of the Higgs boson. It had always been believed that mass was an intrinsic property of a particle – a measure of the "amount of substance" in a particle. However, with the discovery of the Higgs boson it became clear that mass emerged as a result of the interactions between elementary particles and the Higgs field. In other

----

[15] *The case for background independence* by Lee Smolin: http://arxiv.org/abs/hep-th/0507235

words, mass arose from the relationships **between** particles – it is not an inherent property of an elementary particle.

Particle types

On the basis of this discussion of particle properties, you might well feel like raising an objection. How can the properties of particles be determined by other objects in the universe when it is well-known that particles come in different types? For example, some particles are electrons, some particles are photons. Electrons appear to have completely different properties to photons. In that case, you might argue, how can properties not be inherent in a particle?

In order to answer this question, let us examine how different types of particles are supposed to have been formed. This actually raises quite a philosophical question: why is there any structure at all in the universe? Why should the universe not be perfectly homogeneous, entirely composed of just one substance, or one type of particle? In such a situation, the universe would be perfectly symmetrical. Why should this symmetry ever be broken?

In turns out that in the high energies found in the early moments after the big bang, the universe was, indeed, highly symmetrical. Instead of the four fundamental forces we find today, it is believed there was only a single force. At high energies in the universe, such as occurred soon after the big bang, the energy field at each point in space was highly regular. We can illustrate this situation by considering the highly-regular energy at each point in space to be denoted by a pencil balanced on its tip – see Figure a) opposite. The pencil in this position possesses rotational symmetry – it appears the same when viewed in all horizontal directions along the table. It is a highly-regular orientation.

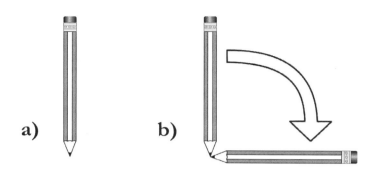

However, this balanced orientation of the pencil is highly unstable. As the energy of the universe drops, the pencil falls flat, as shown in Figure b).

As a result, the symmetry of the energy field is broken – the pencil no longer appears the same in all horizontal directions as we look along the table. We have moved from a symmetric situation to an asymmetric situation. This energy field is capable of producing different types of particles with different, asymmetrical properties. We therefore find groups of different types of particles – such as electrons and photons – being produced by symmetry breaking.

So does this mean different particle types have different properties? If so, then it would appear to contradict the principle that particle properties only arise through interaction with the rest of the universe. Indeed, it would appear to imply that a particle has properties which are absolute, which are inherent to the particle, and would still exist if the particle was the only thing in existence, completely isolated from the rest of the universe. If so, then that would throw our whole proposed theory into doubt.

Fortunately, this is not the case. Yes, particles form groups with the same characteristics. But to say a particle is a member of a group, i.e., to say a particle is of the same type as another particle, is a **relative** measure. It is just a relative comparison. Once particles are categorised into groups it is then the properties of the entire **group** which is determined by the rest of the universe. We might identify a group of particles in the universe which share the same characteristics and we might call those particles "electrons". However, the properties of the particles in that group of electrons would be determined by their relationship with the rest of the universe. Indeed, it would be determined by its relationship with other particle groups – the group of photons, for example.

Particle properties are always relative.

The quantum rationale

We now have a principle that particle properties are defined solely by the relationships between the particle and the rest of the universe. Now, by turning this principle on its head, we are led to a remarkable and important conclusion. If the rest of the universe is responsible for defining the properties of a particle, then **an object which is isolated from the rest of the universe must have completely undefined properties**. There would be no way for Nature to assign any property values – absolute or relative – to that object. The property values would be fundamentally undefined.

Let us reconsider a particle which is isolated from the rest of the universe, or has been generated as a new particle and has not yet interacted with the rest of the universe, i.e., it has not yet been measured or observed. What can we say about its properties? Once again, in the absence of absolutes, Nature is fundamentally unable to assign any absolute values to the particle. And, because the particle has not yet interacted with the rest of the universe to be measured or observed, Nature is unable to assign any relative values either. As discussed in the last section, all we can say about its property values is "They go up to 11", an essentially meaningless statement reflecting the fact that the particle's property values are undefined – they have no relation to the rest of the universe.

Before observation or measurement, the object must be like a blank sheet: it must be an undefined object with the potential to take **any** possible property values. So, before observation, the object must have a multi-valued form of reality – as is observed in quantum mechanical behaviour. It is only when the object interacts with the rest of the universe that its properties become progressively tied down to particular values.

This is precisely what we saw in the case of environmental decoherence. It is through interaction with the rest of the universe (via the measuring apparatus) that the properties of a particle become fixed. Before observation, the particle's properties are undefined.

So here, rather wonderfully, we have a rationale – an explanation – for the multi-valued nature of quantum behaviour before measurement (or observation). Here are the logical steps we followed to get to this conclusion:

1) Nature has no access to absolutes, so it is always fundamentally unable to assign absolute property values to a particle.

2) Therefore, particle properties are defined by the relationship of that particle with the rest of the universe.

3) Bearing this in mind, if we consider an isolated particle, or a newly-generated particle which has not yet been measured or observed, Nature has fundamentally no way of assigning any form of property values – absolute or relative – to the particle.

4) The properties of the particle are therefore fundamentally undefined – like a blank sheet. Its properties must have the potential to be any possible value.

5) The object must therefore have a multi-valued form of reality before it is observed. It is only after observation – when the object interacts with the rest of the universe – that the properties of the object become fixed.

So we now have an explanation for the apparently peculiar superposition state of objects before observation, an attempt to answer John Wheeler's question "How come the quantum?" We see that this behaviour is not so strange after all – it is, in fact, just what we should expect in a relational universe, in which Nature has no access to absolutes. And because we have come to this conclusion using a logical approach, we should expect this to be true in all possible universes: quantum mechanical behaviour is fundamental.

There have been many descriptions of the multi-valued nature of quantum reality, but these have always been presented as a *fait accompli* – an established fact, which should just be accepted at face value. For example, you may have read statements such as Murray Gell-Mann's *Totalitarian Principle*: "Everything which is not forbidden is compulsory". This describes quantum mechanical behaviour by saying that anything which is not forbidden can and must happen. However, statements like these do not explain **why** we see multi-valued behaviour. The rationale for multi-valued behaviour which has just been presented in this book is the first logical rationale for quantum behaviour I have read. It does not just seek to **describe** quantum mechanical behaviour, it seeks to **explain** it. It is seen that quantum mechanical behaviour is expected behaviour in a relative universe.

In the macroscopic, human world, we do not see quantum mechanical behaviour. Environmental decoherence has long since removed the quantum weirdness before it reaches the macroscopic scale. Just like the roulette wheel at the quantum casino, if you spin the wheel enough times (i.e., make many quantum measurements), the randomness inherent in quantum behaviour becomes unnoticeable. Nature always does the best it can with the tools available, and it generally does a very good job in hiding quantum mechanical behaviour from our eyes. However, when Nature is pushed to its limits, in single measurements at the smallest scales, the lack of absolutes starts to show through and we perceive the inevitable "glitches".

So, once again, we find bizarre behaviour in Nature arising as a **side effect** caused by the lack of absolutes. These side effects only start to show through at the extremes. In our discussion on relativity, we have seen these side effects emerge at extremes of speed, or in the presence of extremely large masses. We now see side effects occurring at extremely small scales.

Quantum magic

In the introduction to quantum mechanics in Chapter Six, I made it clear that I wanted to avoid emphasising the more bizarre aspects of quantum mechanical behaviour, and I hope I was successful in that respect. However, it is true that there are many aspects of quantum behaviour which do appear extremely counter-intuitive and, indeed, bizarre.

We have already considered the double-slit experiment in which a particle can apparently be in two places at one. It is also true that particles can spontaneously appear and disappear out of nothing – like a rabbit pulled out of a magician's hat. Now that we are armed with a greater understanding of quantum behaviour, can we start to make more sense of these impressive tricks from the quantum magician?

The double slit experiment is a classic example of the multi-valued nature of quantum reality, the particle being able to pass through both slits. In the discussion in this chapter, we have seen that this multi-valued behaviour is a consequence of the relative universe is which every particle must be defined in terms of its relations with the other particles of the universe. We naturally think of particle properties as being inherent in a particle, as though the particle would still have those properties if it was the only isolated object in the universe. However, when we change our mindset to accept the idea that particle properties only arise in relation to other particles, we find quantum behaviour begins to make a lot more sense, and we can then start to discover the secrets behind the quantum magician's tricks.

However, it is very hard to get out of the mindset that particle properties are not inherent in the particle. This is why quantum behaviour appears so mystifying. It is a principle employed by many human magicians. For example, the trick in which a rabbit is pulled from the magician's hat appears completely mystifying because you considered the hat to be an isolated object, with its own inherent properties. The magician did his best to divert your attention from the other objects around the hat, such as the table with the hole drilled into it through with the rabbit was passed into the hat. It is only by considering the hat's relation with the other objects in the environment that the mystery is solved – just as the effect of environmental decoherence is a solution to the puzzling quantum measurement problem.

Now let us perform a thought experiment to uncover the secrets of the quantum magician, and see how he manages to pull his particles out of a magic hat. Let us imagine – once again – that we want to measure the length of a plank of wood. Clearly, in order to obtain this measurement, we can only use objects already present inside the universe. This means we can only use the rulers, gauges, etc. which exist inside the universe. There is no alternative way to obtain this reading. There is no absolute axis of length outside the universe which we could use to obtain an absolute, unequivocal reading.

So let us say we use our ruler to measure the object, and, according to the ruler, the object is one metre long. Now, bear with me on this, let's presume something really strange happens. Let's imagine every object in existence in our peculiar universe – apart from our plank of wood – bizarrely shrinks by 50%. We now find that our plank of wood has become two metres long – according to our ruler.

But this is to be expected, you might argue. There is nothing surprising here, you might say. The object has not really doubled in size, you might argue, it is merely our ruler which is giving an incorrect reading. The object was previously measured as being one metre long and, in the absence of any external influence acting on the object, it is always going to be one metre long. The property of being "one metre long" is an intrinsic, unvarying property of the object. At least, that is what you might argue.

Well, you would be wrong. Remember that any object in the universe can only be defined in terms of every other object in the universe – Nature has no other way of defining the properties of an object. There is no absolute axis of size outside the universe to which Nature can refer. If the universe tells us the object has doubled in size then, **by definition, the object really has doubled in size!**

This is extraordinary, and we are left with an apparently bizarre conclusion which goes against all intuition. No force has acted on the object, and yet it has doubled in size. It is as though some self-contained internal magic of the object has made it change its state. It appears a property of the object – which we believed to be innate and unvarying – has been radically modified. We would naturally attribute this change in the object's state to some peculiar mechanism **inside** the object. Just as with the magician's hat, we would think the hat was magic, and try to understand what was special about the hat.

However, the secret to the trick does not lie with the hat, and it does not lie with the isolated particle. The secret to the trick is the environment: the hole drilled into the magician's desk, or the doubling of the scale of the universe. Even the very existence of the particle itself is dependent on the universe around it – if the universe says the particle does not exist then, by definition, it ceases to exist. There is no other external standard by which Nature can determine the existence of the particle.

It is all too easy to ignore the environment and concentrate on the single particle under observation, but it is only by considering the complete magician's set-up, and the relationship of the particle with the environment, that we can finally uncover the secrets of quantum magic.

Symmetry and quantum mechanics

To end this chapter's discussion on quantum mechanics, it is interesting to consider underlying connections between symmetry and quantum mechanics.

It is important to stress that it is not only measurements of space or time that must be taken as relative. As there is a complete absence of any absolutes – of any form – in the universe, this means that Nature cannot have access to absolute standards for **any** type of properties. This explains why the quantum formulation applies to all of Nature, describing all forms of particle properties, not just those associated with space and time.

Carrying on this thought, in our discussion on symmetry in Chapter Three we saw how the lack of absolutes available to Nature caused symmetries. As was explained then, symmetry reflects Nature's fundamental inability to distinguish between one physical situation and another. If Nature is unable to distinguish between two situations, then we can transform one physical arrangement into another without altering the situation. This represents a symmetry.

So if Nature has no access to absolute standards for particle properties, then we would expect to see similar symmetries in particle properties – and this is exactly what we do see. It has long been known that symmetry plays a remarkable role in particle physics. Different types of particles can be classified into groups, those groups exhibiting symmetries. The discovery of new particles has even been predicted because they are required to complete a perfect symmetry. From our discussion, it is clear that the cause of all this symmetry in particle properties is a result of the fundamental limitations on Nature due to the lack of absolutes, which results in Nature's inability to distinguish situations.

In Chapters Three and Four we considered relativity and we found that there was symmetry in the universe in both space and time (space and time translation invariance) due to the lack of absolute space or time. In this chapter we have considered the properties of the elementary particles and, again, we have found symmetry in their properties. On the basis of our analysis we can see that there is the same underlying cause for both types of symmetry: no absolutes. This is a quite remarkable conclusion: an apparent unification of symmetries! Symmetries in space, time, and particle properties have the same underlying cause.

Now let us examine an example of particle property symmetry, the symmetry between positive electric charge and negative electric charge, The technical name for this symmetry is U(1) symmetry:

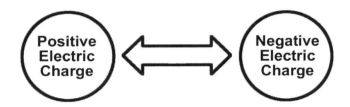

This symmetry arises because the property of positive charge is **defined** as the opposite of negative charge. The two properties – positive charge and negative charge – are defined in terms of each other, and in **no other way**. This is perhaps what we might have expected to find based on our discussion throughout this book. There is no absolute definition of any particle property: particle properties **must** be defined relative to each other.

Because these properties are defined solely in relative terms (and not in absolute terms), if every positively charged particle in the universe became negatively charged, and simultaneously every negatively charged particle became positively charged, then the overall situation would be unchanged – the universe would continue as if nothing had happened. This is because the relation between the charged particles (which is the only important measure) would be unchanged.

This means that we are essentially free to define "positive charge" differently at every point in space. However, Nature needs to have some way of ensuring that "positive charge" at one point in space is the same type of charge as "positive charge" at a different point in space – it needs to keep track throughout the universe. Otherwise, particle charges would be free to swap between positive and negative at random, and the result would be chaos (and would also break the law of conservation of energy [16]). The way Nature keeps track is to have a field passing through space, defining which type of charge is actually the definitive positive charge. This type of field is called a *gauge field*, and is a result of the perfect symmetry of elementary particles. For the case of electric charge, the gauge field is the electromagnetic field and is carried by photons.

To quote Robert Oerter from his book *The Theory of Almost Everything*: "Our inability to define the charge in any **absolute** [my emphasis] sense is the ignorance that gives rise to a quantum field: the photon field." So not only is particle symmetry due to Nature's inability to access absolutes, we now find that the electromagnetic field is also due to this restriction on Nature. The electromagnetic field allows Nature to keep track of charge in the same way that the gravitational field allows Nature to keep track of masses (as per the discussion in Chapter Three). In fact, it can be shown that all four fundamental forces (electromagnetic, strong, weak, and gravity) are a direct result of types of gauge symmetry.

This is actually another example of quantum mechanical behaviour. We note that the type of electric charge (positive or negative) is only defined relative to the other type of charge – it is not defined absolutely. So an isolated charge has to be considered as essentially multi-valued: it has the potential to be either positively or negatively charged. It is only through interaction with the rest of the universe – via the gauge field – that the type of charge becomes fixed. So once again we see quantum mechanical behaviour: before observation an isolated particle appears to be multi-valued. Only when it makes contact with the environment does it get

---

[16] For details, see Chapter Eight of Bruce Schumm's book on particle physics: *Deep Down Things*.

tied-down to a particular value from the many potential values.

In this chapter, the hypothesis has been presented that the root cause of quantum mechanical behaviour is Nature's fundamental inability to access absolutes. Whatever doubt might remain about this has surely been eliminated by our study of the quantum mechanical behaviour of gauge fields. The earlier quote from Robert Oerter makes this abundantly clear. According to Oerter, the electric charge is initially in a multi-valued superposition (i.e., quantum) state due to Nature's "fundamental inability to define the charge in any **absolute** sense". So here is an example of quantum mechanical behaviour which is definitely due to a lack of absolutes. The orthodox, established science of gauge theory provides us with final confirmation that the root cause of quantum mechanics is Nature's fundamental inability to access absolutes.

What is more, we will see in the next chapter that this is the same fundamental cause which links relativity with quantum mechanics.

9

HIDDEN IN PLAIN SIGHT

We now move on to consider the fundamental link between relativity and quantum mechanics. And, I am pleased to say, as we have covered so much background already, the derivation of our result is trivially simple.

It has emerged that the counter-intuitive "glitches" of both quantum mechanics and relativity are **side effects**, side effects caused by Nature's inevitable switch to a relative mode of operation. This is inevitable due to Nature's complete inability to access any absolutes in the universe. This is due to there being no absolute axes – no absolute standards of reference – of any kind outside the universe (remember: "there is nothing outside the universe").

This relative universe inevitably leads to the theory of relativity, in which no observer is preferred. Also, in a relative universe, it is inevitable that particle properties can only be defined relative to other particles. This implies that an isolated particle, or a newly-generated particle before observation, is in an undefined, potentially multi-valued state. This inevitably leads to what we view as quantum mechanical behaviour: another "glitch", another side effect.

The following diagram shows the path of logical deduction, starting from our fundamental principle, and leading inevitably to both quantum mechanics and relativity:

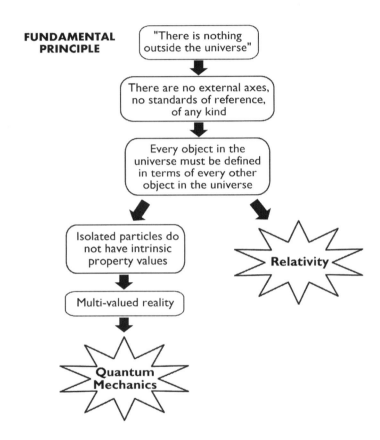

We do not notice either quantum mechanical behaviour or relativistic behaviour at human scales and speeds (again, this is a good clue that both theories have the same underlying cause). At human scales and speeds, we have access to absolute measurements, because we can define absolute standards of measurement outside of a closed system. We have access to measuring tapes which provide us with absolute measurements of length, and speedometers which provide us with absolute measurements of speed. These can be related – via units – to some absolute standard of measurement (remember the platinum bar in Paris?).

However, Nature has no such luxury of access to absolutes. There can never be an absolute standard outside of Nature's closed system (i.e., the universe). When Nature is pushed to the to the extremes, at the most fundamental levels, all of the absolutes get progressively stripped away. This is the point when we start to see the "glitches" which are caused by the Nature's switch to relative measures.

Nature always tries to keep the universe running smoothly by doing the best it can using the limited tools at its disposal (i.e., having to use relative measurements). For the most part, Nature does a very good job and it appears as though Nature has access to absolutes. However, the side effects start to show through at the extremes: at extremes of speed, in the presence of extremely large masses, or at extremely small scales.

The table on the next page shows the theories of quantum mechanics and relativity, revealing that both mechanisms have the same underlying cause (you will see the entries in the first column entitled "Underlying Cause" are the same for quantum mechanics and relativity). The table also shows the limitations on Nature caused by the inevitable shift to a relative mode of operation, and reveals the subsequent "glitches" in behaviour which we interpret as quantum mechanical or relativistic behaviour:

Theory	Underlying Cause	Nature's Limitation	Nature's Best Effort	The Side Effect
Quantum Mechanics	Nature has no access to absolutes – has to switch to a relative mode of operation.	Object properties have to be defined **relative** to all other objects in the universe.	No multi-valued behaviour noticeable at human scales.	Multi-valued quantum mechanical behaviour before observation.
Special Relativity	Nature has no access to absolutes – has to switch to a relative mode of operation.	No preferred observer – speed of light is constant **relative** to every observer.	No time dilation noticeable at human speeds.	Time dilation at high speeds.
General Relativity	Nature has no access to absolutes – has to switch to a relative mode of operation.	No absolute space – gravitational field must be defined **relative** to the masses in the universe.	Space appears flat in absense of huge masses.	Curvature of space in presence of huge masses: gravity.

It is truly surprising that this link between quantum mechanics and relativity appears to have passed unnoticed. As we have discussed, it has long been known that relativity is due to the absence of **absolute** space, and, as we have found in the quote from Robert Oerter, it has been known that the quantum mechanical behaviour of gauge theory has been due to Nature's "fundamental inability to define the charge in any **absolute** sense". So here we have a quote describing the cause of relativity as being due to a lack of absolutes, and another quote describing the cause of quantum mechanical behaviour as also being due to a lack of absolutes. So why has this obvious connection between quantum mechanics and relativity not been identified earlier? It is not as though this is arcane knowledge hidden away in obscure scientific journals. Quite the opposite in fact: both of these quotes are taken from popular science books which you could find in your local bookstore! I am quite staggered that such an obvious link could have been overlooked for so long.

Observer dependence

Can this approach lead us to find more fundamental connections between quantum mechanics and relativity?

Well, in our study of quantum mechanics in Chapter Six, we considered the importance of the role of the observer in quantum mechanics. The role was of such significance that, at one time, it was even believed that a human consciousness was required to bring reality into existence at the quantum mechanical level. If quantum mechanics and relativity truly share a common root – as proposed in this book – we would expect relativity to exhibit a similar fundamental observer dependence. However, at first glance, it would appear this crucial observer dependence is not found in relativity. This belief is expressed in a quote from Lee Smolin in his book *Three Roads to Quantum Gravity* comparing Newton's view of the observer to Einstein's view: "The problem is that, while quantum theory changed radically the assumptions about the relationship between the observer and the observed, it accepted without alteration Newton's old answer to the question of what space and time are. Just the opposite happened with Einstein's general relativity theory, in which the concept of space and time was radically changed, while **Newton's view of the relationship between observer and observed was retained.**"

However, the more you think about it, the more you realise that "Newton's view of the relationship between observer and observed" was most certainly **not** retained in the theory of relativity. In fact, relativity is the very epitome of an observer-dependent theory!

Let's start by asking a simple question: how fast is the spaceship travelling in the image below?

Clearly, if there is "nothing outside the universe" then, in the absence of any absolute measurement scale outside the universe, we cannot assign any absolute speed to the ship – we have nothing to measure it against. We cannot assign a speed to the spaceship on which all observers in the universe will agree. In fact, the only way we can assign any numerical value to the speed of the ship is by giving up all hope on finding an absolute, observer-independent speed, and by considering its relative speed instead.

So, if we want to obtain a numerical value for the speed of the spaceship, we have to first define our observer:

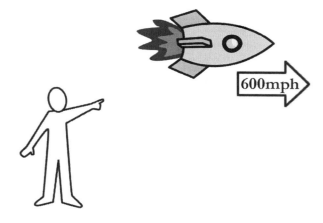

In the image above, we have defined a position for our observer and we can now obtain a value for the velocity of the spaceship – relative to our observer – which happens to be 600mph, travelling to the right.

However, according to Newton, the function of the observer was unimportant. In Newton's absolute space, it can be stated with certainty that the spaceship is moving at an absolute speed of 600 mph. The role of the observer plays no part in determining the velocity of the spaceship. However, when the theory of relativity was proposed, the idea of absolute space was refuted. Let's say we make contact with the captain of the spaceship and ask him to "observe our observer". According to the captain, it is in fact our observer who is moving at 600mph, and, as no frame of reference is preferred in relative space, the captain must be considered to be just as correct as our initial observer. In relative space, if a spaceship is moving at 600mph relative to an observer, it is just as valid to say the observer is now moving at 600mph relative to the spaceship. In Newton's model, the observer is stationary, but in Einstein's model it is valid to say the observer has a relative speed of 600mph! To answer Lee Smolin, I would say that constitutes a rather severe modification of "Newton's view of the relationship between observer and observed"!

As I said earlier, relativity is the very epitome of an observer-dependent theory. And this point is vital to this current discussion: quantum mechanics and relativity are both observer-dependent theories.

Now let us imagine we have a second observer on a second spaceship travelling to the left at 400mph:

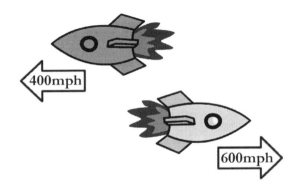

According to this second observer, the first spaceship is in fact travelling to the right with a relative velocity of 1000mph (the sum of 600mph + 400mph). So we now have two different observers who make two different measurements for the velocity of the spaceship. Our initial observer measures a value of 600mph, whereas our second observer (in the second spaceship) measures a value of 1000mph. This clearly shows how observer-dependent this velocity measurement really is.

We have seen that there can be no measurement of velocity without first defining an observer. To assign any form of velocity to an isolated object with no relation to the rest of the universe would be absolutely meaningless. Velocity is only defined relative to another object in the universe.

At this point, alarm bells should be ringing in your head – where have we heard a similar statement before? It is exactly the same principle as we revealed in the discussion of particle properties in the previous chapter: the properties of a particle are determined by its relationship with all the other objects in the universe. So now have uncovered the same principle underlying the properties of a particle and the measurement of relative velocity. Could this possibly be starting to sound like the start of another link between quantum mechanics and relativity?

But in our discussion of particle properties in the last chapter, we derived the result that objects which were isolated, or had never been measured, would have properties which would have to be in an undefined state, a strange multi-valued state. This led us to form a rationale for quantum mechanical behaviour. But this is surely not true for the spaceship's velocity – the spaceship's velocity is clearly not in a multi-valued state before observation.

Or is it?

As we have just seen in our previous illustration of measuring the spaceship's velocity, one observer measured the velocity of the spaceship at 600mph, while the other observer measured the velocity of the spaceship at 1,000mph – clearly a multi-valued reality! In the absence of absolute space, either velocity value – or both – is equally valid.

But having two different answers to our question "How fast is the spaceship travelling?" is not a very satisfactory answer. Can't we do any better? Can't we get an observer-independent, absolute value for the velocity of the spaceship – before any observer-dependent measurement is taken? Well, the standard answer is, no, in our universe only relative velocity has a meaning. But, you might say, that is totally unsatisfactory. Before any measurement is taken, the spaceship clearly has a velocity – we know it is moving. So why can't we determine its value before observation? Its velocity clearly has a reality before observation – why can't we assign a value to it?

Well, I suppose the best we could do is assign the spaceship with a **potential** velocity which has to take **any possible value** from zero up to the speed of light. For example, if the spaceship of the second observer in the previous example is travelling at a million miles an hour, we now find our measurement of the initial spaceship's relative speed is 1,000,600mph. With an infinity of potential observers in the universe, we could say that the before-measurement, observer-independent velocity of the spaceship is **any** possible value.

So let's recap:

1) Before measurement, we know the spaceship's velocity has a reality (we know the spaceship is moving), but we cannot assign a value to it.

2) If we want to obtain a measurement then we have to specify an observer as the measurement is completely observer-dependent.

3) But if we want to assign a velocity to the spaceship **before** observation we have to assign all possible velocity values to it.

So where have we heard a description like this before? **There are clear parallels with quantum mechanics!**

Compare this with the situation of quantum mechanics:

1) In quantum mechanics, we know a particle has a reality before observation, but we cannot assign a value to its properties.

2) If we want to obtain a measurement then we have to specify an observer as the value is completely observer-dependent.

3) But if we want to assign a value to its properties **before** observation, we have to assign all possible values to it.

These three points are absolutely identical to the three points raised in our relative velocity scenario!

So clearly the two scenarios – relative velocity (which leads to relativity) and quantum mechanics – show astounding similarities! This is because quantum mechanics and relativity share the same underlying cause, a common root to their behaviour. And that common cause is a universe in which Nature is denied access to absolutes, a universe is which everything must be defined in terms of everything else, and in which all of reality is relative.

It might seem rather astonishing that no one has pointed out this incredibly simple link before. A startlingly obvious link between relativity and quantum mechanics due to the two theories having the same underlying cause. I believe this has gone unnoticed purely due to the quite stupendous simplicity of the solution – it has truly been hidden in plain sight.

Actually, when I say no one has pointed this out before, I must say that the loop quantum gravity researcher Carlo Rovelli has hinted at a connection: "Quantum state and values that an observable takes are relational notions, in the same sense in which velocity is relational in classical mechanics (it is a relation between two systems, not a property of a single system). I find the consonance between this relationalism in quantum mechanics and the relationalism in general relativity quite striking. It is tempting to speculate that they are related."[17]

Fundamentally identical behaviour

I believe it has been shown that the underlying principle behind quantum mechanics and relativity are the same: a relative universe in which Nature – unable to access absolutes – is forced to switch to a relative mode of operation. Additionally, I believe that because we built-up to this conclusion from fundamental principles, then this behaviour is fundamental behaviour which would necessarily be true in all conceivable universes.

Although there remains some clear differences between quantum mechanics and relativity, I do not believe these are **fundamental** differences. These differences should not hide the fact that the two theories are **fundamentally** identical.

Perhaps the most obvious difference is that quantum mechanical observations appear to have an irreversible effect on the object under observation, as was discussed in Chapter Seven, whereas there appears to be nothing irreversible about relativity. We only ever see quantum jumps in the forward time direction, from a superposition state to a well-defined state – never vice versa. However, we will see that this irreversible effect does not reflect a fundamental difference between the two theories.

---

[17] *Quantum spacetime: what do we know?* by Carlo Rovelli: http://arxiv.org/abs/gr–qc/9903045

CONCLUSION

Firstly, the vastly greater effect on the observed object in quantum mechanics can be explained in terms of the scale of the observation. In a measurement of relativistic speed, for example, if a human observer measures the velocity of a spaceship, he need only bounce a few photons off the spaceship – not enough to alter the course of the spaceship in any measurable way. However, in quantum mechanics, the effect of the observation on the observed object can be overwhelming. This is because the scale of the object being observed (a particle) is of the same order of magnitude as the particle performing the observation.

Moving on to consider the irreversibility of the measurement, this can be explained in terms of increasing entropy – due to the second law of thermodynamics – during the process of environmental decoherence. Just like the irreversibility of a wine glass breaking in the forward time direction, we find irreversible quantum jumping in the forward time direction: the irreversibility of quantum mechanics is due to increasing entropy.

So we should now realise that this principle of increasing entropy also applies to relativistic measurements. Bouncing a few photons off a spaceship to measure its speed inevitably affects the object under observation – no matter how minuscule the effect. This inevitably increases the entropy of the system (due to the second law of thermodynamics) and so is also an irreversible process. We now see that **both** quantum mechanics and relativity are irreversible. It is one more way in which the two theories – previously considered to be completely distinct – turn out to be fundamentally identical.

Quantum mechanics and relativity share the same fundamental cause. The only differences between the two effects are caused by the scale of the measurement. These two great theories are simply two different manifestations of the same fundamental principle.

10

CONCLUSION

In this book, an incredibly simple link has been suggested between relativity and quantum mechanics. By building-up from fundamental principles, it can be shown that both theories share a common root: the fundamental inability of Nature to access any absolutes in the universe.

It is always unattractive to oversell one's hypotheses, which I why I generally use the term "link" rather than "unification" to describe this relation between quantum mechanics and relativity. Indeed, I would imagine most theorists would dismiss this idea of a potential link as representing any kind of unification. It certainly does not resemble the attempts currently being made in string theory and loop quantum gravity. It appears too absurdly simple. It contains no equations, or new, testable predictions.

However, if pushed, I would suggest that this does indeed represent a unification. It contains no equations because it is a unification *in principle*, in much the same way that several unifications – such as Galilean equivalence – are described by a principle rather an equation. It might be argued that a unification which contains no equations is of little practical use, but I do not see criticism of Galilean equivalence on that basis. Indeed, Galileo's principle was also a result of Nature's fundamental inability to access absolutes. It would appear there is a common theme here.

The hypothesis described in this book was always going to differ from the conventional results of quantum gravity research. Conventional research works from the "top down" in that it attempts to mesh together quantum mechanics and relativity. This approach usually involves the application of random quantum behaviour to the smooth structure of space. Instead, the methodology of this book has been "bottom up" – attempting to build from fundamental principles. Unlike conventional approaches attempting to mesh relativity with quantum mechanics, it does not just seek to **describe** the effects of the unification – it seeks to **explain** it.

And this, perhaps, provides us with the reason why such a simple solution to the problem of unification could have gone unnoticed for so long: the proposal depends on an explanation of the foundations of quantum mechanics. Sadly, with a few exceptions, there is little interest or research in foundational questions of quantum mechanics at the moment. This has not always been the case – the rapid progress in quantum mechanics in the early years of the twentieth century placed foundational questions at the forefront of research. However, in the current academic climate, foundational questions seem to be considered the remit of philosophy – not physics, and get precious little attention. As a relevant example, string theory and loop quantum gravity seem to have nothing to say about the quantum measurement problem – an absolutely fundamental component of quantum behaviour.

This is a great shame, because if the principle described in this book is even remotely accurate, it would appear that foundational questions have a vital role to play in uncovering the unification of relativity and quantum mechanics. It would appear true unification can only be found at a deeper level of analysis than is considered by current approaches which attempt to mesh relativity and quantum mechanics together at a higher level of reality. It is not my fault if people are looking for unification at the wrong level. If the work described in this book is correct, then until foundational questions are tackled head-on, there will be no unification.

This conventional approach to unification – of attempting to mesh relativity and quantum mechanics together at a high level – leaves me with a rather surprising thought. Surely, the first approach one should try when attempting to unify two theories should be to stand back and consider the **similarities** between the two theories. What do the theories have in common? This would seem the obvious approach. The commonality would then direct the researcher towards the unification. I like to think that this approach has been followed in this book, especially in the common link revealed in Chapter Nine. However, it seems this "obvious approach" of considering the similarities between the theories is not what has happened. Instead, there appears to have been something of a "brute force" approach to unification, attempting to weld the theories together. As far as I am aware, no one has bothered to stand back and say "What have these theories got in common?"

To those who would criticise the hypothesis described in this book on the basis that it is absurdly simple, I would simply direct them to the discussion in Chapter One which revealed how all unifications in physics have been based on incredibly simple ideas, ideas which can be understood by anyone. Indeed, based on the discussion in Chapter One, it was stated that any successful unification hypothesis **has** to be extremely simple – a criterion which is not met by string theory or loop quantum gravity. To criticise a unification hypothesis on the basis of it being overly-simplistic seems perverse to say the least!

Certainties and absolutes

I would like to finish the book with a few general discussion points which have been raised by the proposal.

As has been discussed throughout this book, at the human macroscopic scale, we are able to operate as though we have access to absolute scales of measurement. We can access measuring tapes and speedometers which apparently allow us to make absolute measurements of length and speed. However, what must be realised is that in a relative universe everything must be defined in terms of everything else in the universe – there are no genuine absolutes. This inevitably results in very circular definitions of measurements. Just like the example of the apple being held in the hand in Chapter Eight: "I know the apple is real because I can hold it in my hand", we find that the whole of reality is composed of a self-supported structure of relationships. In this case of the apple, we assume an object to be real because of its relationship with another object which we already presumed to be real: the hand.

So all the objects in the universe – at the macroscopic level – are connected in an endlessly complex mesh of interrelationships. Everything supports everything else. This gives us the luxury of an impression of absolutes and certainties: our impression of reality might be an illusion, but that illusion is at least supported by the rest of the universe.

Considering the position of particles, for example, the relative nature of environmental decoherence means we get an impression of complete certainty as to the position of a particle. The environment – the universe around the particle – works together to provide an impression of certainty. However, as we delve deeper into the fundamentals of the universe, we find that self-supporting network is progressively stripped away. Eventually, we reach the fundamental layer when objects can be isolated from the rest of the universe. At this point, we say quantum mechanical effects are dominant, but what we really mean is that that mesh of certainty is completely stripped away. At this point, we see our macroscopic certainties and absolutes for what they really are: an illusion of the relative universe. We find certainties stripped away in the Heisenberg uncertainty principle. We find absolutes stripped away in similar fashion, leading to multi-valued behaviour.

Put simply, there are no such things as certainties and absolutes. In the mathematical world, yes, we can find these things, but not in the physical world.

When Nature is stripped naked, we find the emperor has no clothes (or certainties or absolutes). We call the result quantum mechanics.

> *Quantum physics makes me so happy. It's like looking at the universe naked.*
> – Sheldon Cooper

It is interesting that as we dig ever deeper into the foundations of Nature, and we lose certainties and absolutes, we actually find we encounter more symmetries. As we discussed in Chapter Three, this is because symmetry reflects Nature's fundamental inability to distinguish between one physical situation and another. If Nature is unable to distinguish between two situations, then we can transform one arrangement into another without altering the physical situation. This represents a symmetry.

So at the fundamental levels, the limitations on Nature means we lose certainties and absolutes, but those same limitations mean an increase in fundamental symmetries. Physicists are well aware that the more symmetrical theories tend to represent deeper underlying truths about Nature.

Many Worlds (revisited)

The universe is a structure of such elegance, simplicity, and economy that I can't help thinking that whoever first proposed the concept of multiple parallel universes did Nature a great disservice – that is to ignore the evident simplicity and economy of the universe. I am sure the proponents of parallel universes did it with the best intentions but, really, the damage has been done. It seems almost impossible to pick up a popular science book these days without reading how there exists a parallel universe in which a forty-foot gorilla has just won the Olympic gold medal in synchronised gymnastics.

Articles such as that must make great copy – and maybe good book sales – but the message of physics is one of economy and simplicity. In physics we find energy is minimized, structures are symmetrical, simpler explanations are preferred over more complex explanations. As William of Occam said: "Entities should not be multiplied unnecessarily". We should only add complexity to our explanations as a last resort, if there is no other solution available. At all times, the thrust of our drive should be towards simplification.

Ironically, by postulating the existence of multiple parallel universes, researchers are missing out on the analytical power of the principle that the universe is the one thing that exists. By saying that "everything that can possibly happen will happen", the theorists who advocate parallel universes are losing any constraints to their theories. Anything goes, basically. But, ironically, constraints can make our job much easier. Imagine you are an author, and you are seeking inspiration for a new book. You might sit down in front of your typewriter or word processor and stare blankly at a sheet, waiting for inspiration before you can get started. However, if you are commissioned to write a book about a certain subject – for example, a book about a kidnapping – that instantly gives you a start and gets your mind thinking along certain lines. Constraints help us in finding a solution. In fact, if the problem becomes completely constrained – your commission telling you the entire plot of the book – then the solution becomes completely obvious.

A theory that predicts everything predicts nothing – it is completely unconstrained. This makes our job of finding a solution much harder. By imposing fairly obvious constraints, such as saying there is nothing outside the universe, we make our job of finding a solution much easier, because then we realise that our solution must be found within the universe, and not in some parallel universe scenario. In the case of relativity and quantum mechanics, I believe it is the constraint that there is nothing outside the universe which, indeed, completely constrains the problem so the solution does, indeed, become completely obvious. I believe the results in this book show how the constraint leads directly to both quantum mechanics and relativity. This result could never have been derived if we had lazily jumped to the conclusion that we needed parallel universes to explain these phenomena.

We should not be lazy. We should seek to explain the universe on the basis of what we already know, not on the basis of some flight-of-fancy science fiction. Wild speculation does our quest more harm than good, and hinders our progress. The answers we seek are contained within the universe. The answers we seek are located much closer to home than we might imagine.

Reductionism and emergence

In its efforts to understand the universe, physics has always taken a *reductionist* approach. This means that research has progressed based on the principle that we can understand things by breaking them down – and analyzing – their smallest constituent parts. As an example, particle physics attempts to discover the smallest particles and the most basic forces. The belief is that the smaller the particle, the better your understanding of the physical world.

For many simple applications this approach works well. For example, analyzing the large-scale properties of materials on the basis of their particle chemistry (finding the strength of materials, the temperature at which they melt, etc.). However, for more complex applications this approach proves woefully inadequate. This is because many large-scale phenomena which result from the complex interaction of billions of particles could never be predicted purely by analyzing the properties of a single particle. In this way, the whole is greater than the sum of its parts.

One classic example of an emergent phenomenon is intelligence. You could never realise that such an extraordinary phenomenon could arise if you followed the reductionist approach of analyzing only a single brain neuron. A single neuron is just a simple switch, basically. There is no clue in the structure of a simple switch that it might lead to high-level brain function.

Throughout this book, it has been stressed that the properties of single particles should not be considered in isolation. Properties describe the relationship between objects and arise through the interaction of an object with the rest of the universe. This clearly shows that a reductionist approach is not suitable at the fundamental level.

If, for example, a single particle is moving, how can we define its velocity? Clearly, the velocity is not an inherent property of the particle, but, rather, it has to be defined relative to some other particle. So if we add an additional particle, we can now determine the velocity of the first particle relative to the second particle. But what quantitative result can we possibly get with just two particles? Not very much is the answer – we can just determine that the particles are moving relative to each other. In order to quantitatively measure that velocity, we need some standard measure of speed, and that means we need some more particles with which we can produce standards of distance and time, eventually forming meters and clocks. In fact, to get any useful meaning from our initial particle, we find we need more and more particles, introducing additional layers of meaning. So we see that properties of particles are not held in single particles, but rather those properties progressively **emerge** as the universe is built-up in layer-upon-layer of meaning.

In fact, this shows the importance of treating the universe as one object, with properties of particles only emerging when the universe is treated as a whole.

This realisation raises a problem for physics research which been firmly based on a reductionist model of the universe. The mathematics used by physicists is simply not up to the job of analysing a complex, emergent universe. The neat and simple equations used by physicists might be only suitable for analysing a "toy model" simplified version of the universe, which ignores the possibility of emergence, complexity, and chaos. These processes which can be simplified are called *computationally reducible*. As Stephen Wolfram has said: "Almost all of what traditional equation-based science has been doing is looking just at those computationally reducible parts." However, if the universe is truly computationally **irreducible** then it might be simply impossible to analyse it using simple equations. In fact, the only way to analyse it would be to create a computer with the same complexity as the universe itself!

The laws of Nature

In Chapter One, it was stated that one of the goals of this book was to determine how much of the universe could have been created differently, and how much is a logical necessity. The eventual goal was expressed by Einstein: "What I am really interested in is whether God could have made the world in a different way; that is, whether the necessity of logical simplicity leaves any freedom at all."

Because our approach was to build-up from fundamental principles, we could be sure that any result we discovered would have arisen from logical necessity. Rather astonishingly, we discovered that a remarkable proportion of the laws of Nature could be derived from our fundamental principle – that there is "nothing outside the universe". From the analysis, it appeared that both relativity and quantum mechanics could be derived from the fundamental principle, and so both these behaviours would appear to be fundamental, i.e., there would have to be quantum mechanical and relativistic behaviour in any conceivable universe. Not only that, we find we can derive the requirement for a gravitational field, symmetry in both space and time – and symmetry of particle properties, the requirement that the total energy of the universe is zero, and the necessity of gauge fields. The fact that we can derive such a high proportion of the laws of Nature from our fundamental principle gives us great confidence that it is, indeed, the fundamental principle.

I would imagine that any laws of Nature which could be derived from the fundamental principle – "there is nothing outside the universe" – could be taken to be fundamental laws. In fact, that could be the very definition of "fundamental".

A remarkable proportion of the laws of Nature, therefore, could be considered to be necessary and not contingent. However, it remains true that it is hard to imagine how the arbitrary nature of the physical constants of Nature could ever be derived from fundamental principles. The 19 numerical constants in the standard model of particle physics, for example, appear to have values which are completely arbitrary. It would appear that some form of spontaneous symmetry breaking is responsible for setting the values of the constants, but this still does not explain how the particular values were produced.

To my mind, on the basis of this study, it appears that the major laws of Nature (quantum mechanics, relativity, the second law of thermodynamics, etc.) appear to be fundamental, derived from fundamental principles, and would necessarily have to apply to every conceivable universe. However, it is the remaining arbitrary free parameters which would appear to give the universe its actual form (the properties of the elementary particles, etc.), and it would appear that these values could not be derived from fundamental principles.

This problem is exacerbated by the fact that it appears the fundamental constants have fine-tuned values which have resulted in a "life-friendly" universe. The fashionable answer to this conundrum seems to be to resort to anthropic reasoning, and to suggest the constants are set to different values in different parallel universes. However, the central tenet of this book is that there is only one universe. In the analysis in this book, we have found that building-up from that principle has proved to be a wonderfully effective predictive tool, which suggests that the principle is correct. In which case, to find the solution to the apparent fine-tuning of constants we should restrict our search to solutions contained **within** the universe. I would agree with Lee

Smolin from *Three Roads to Quantum Gravity*: "This first principle means that we take the universe to be, by definition, a closed system. It means that the explanation for anything in the universe can involve only other things that also exist in the universe."

The secrets we seek can all be found **within** the universe. We just have to look more closely – they are hidden in plain sight.

FURTHER READING

www.whatisreality.co.uk
My website

Three Roads to Quantum Gravity by Lee Smolin
Fascinating guide to the state-of-the-art, by one of the leading researchers in the field. Highly-relevant to this book.

The Theory of Almost Everything by Robert Oerter
A comprehensive and readable account of the development of the standard model of particle physics, including a useful exploration of symmetry.

Quantum by Jim Al-Khalili
The best introduction to quantum theory.

The Elegant Universe by Brian Greene
A thorough guide to string theory, unification, and quantum gravity.

The Fabric of the Cosmos by Brian Greene
Excellent guide to relativity, Mach's principle, and the gravitational field.

Introducing Time by Craig Callender and Ralph Edney
A guide to time, and the history and principles behind the block universe theory.

The Complete Idiot's Guide to String Theory by George Musser
Not really a guide to string theory – more a comprehensive and accessible introduction to all aspects of quantum gravity research.

Simply Einstein by Richard Wolfson
A comprehensive walkthrough of relativity.

The Mind of God by Paul Davies
Fundamental questions about why the universe is the way it is. Could the universe have been built differently?

The Trouble with Physics by Lee Smolin
A provocative and informative book.

Deep Down Things by Bruce A. Schumm
An excellent guide to particle physics which is more technical than Robert Oerter's book.

Introducing Quantum Theory by J.P. McEvoy and Oscar Zarate
The "Introducing" series of books are extremely well constructed.

New Theories of Everything by John D. Barrow
Deep consideration of the laws of Nature.

The Fabric of Reality by David Deutsch
Fascinating discussion of reality and the block universe model.

ACKNOWLEDGEMENTS

Several people have been a considerable annoyance during the writing of this book. Chris Parker and Steve Skelton have repeatedly pressed me to include teenage vampire love interest, while Paul Noonan's suggestion of offensive chapter titles has also been incredibly irritating.

However, I would like to thank Chris for all his help with the graphical layout, general comments and ideas, and proof-reading, and Paul for additional proof-reading.

Steve, unfortunately, was of little use.

I have to thank Joseph Wouk for encouraging me to write this book, and I have to thank everyone who has posted comments on my website over the years.

2823906R00113

Printed in Great Britain
by Amazon.co.uk, Ltd.,
Marston Gate.